Teaching Resources

Grade 1

Harcourt Brace & Company

Orlando • Atlanta • Austin • Boston • San Francisco • Chicago • Dallas • New York • Toronto • London

http://www.hbschool.com

ISBN 0-15-311116-X

6 7 8 9 10 085 03 02 01

CONTENTS

▶ Number and Operations

▶ Money

▶ Time

TEACHING RESOURCES

Various types of resources for lessons and practice activities in the Teacher's Edition are included in this section. Some of these resources may also be used with the Learning Center Cards for this grade level.

The resources are provided for the following categories:

- ▶ **Number and Operations**
- ▶ **Money**
- ▶ **Time**
- ▶ **Measurement**
- ▶ **Geometry**
- ▶ **Data and Graphing**
- ▶ **Workmats**
- ▶ **Other**

$$\begin{array}{r} 0 \\ +1 \\ \hline \end{array}$$

$$\begin{array}{r} 0 \\ +2 \\ \hline \end{array}$$

$$\begin{array}{r} 0 \\ +3 \\ \hline \end{array}$$

$$\begin{array}{r} 0 \\ +4 \\ \hline \end{array}$$

$$\begin{array}{r} 0 \\ +5 \\ \hline \end{array}$$

$$\begin{array}{r} 0 \\ +6 \\ \hline \end{array}$$

$$\begin{array}{r} 0 \\ +7 \\ \hline \end{array}$$

$$\begin{array}{r} 0 \\ +8 \\ \hline \end{array}$$

$$\begin{array}{r} 0 \\ +9 \\ \hline \end{array}$$

1 +0	1 +1	1 +2
1 +3	1 +4	1 +5
1 +6	1 +7	1 +8

$\begin{array}{r} 2 \\ +0 \\ \hline \end{array}$	$\begin{array}{r} 2 \\ +1 \\ \hline \end{array}$	$\begin{array}{r} 2 \\ +2 \\ \hline \end{array}$
$\begin{array}{r} 2 \\ +3 \\ \hline \end{array}$	$\begin{array}{r} 2 \\ +4 \\ \hline \end{array}$	$\begin{array}{r} 2 \\ +5 \\ \hline \end{array}$
$\begin{array}{r} 2 \\ +6 \\ \hline \end{array}$	$\begin{array}{r} 2 \\ +7 \\ \hline \end{array}$	$\begin{array}{r} 3 \\ +0 \\ \hline \end{array}$

Addition Fact Cards

$\begin{array}{r} 3 \\ +1 \\ \hline \end{array}$	$\begin{array}{r} 3 \\ +2 \\ \hline \end{array}$	$\begin{array}{r} 3 \\ +3 \\ \hline \end{array}$
$\begin{array}{r} 3 \\ +4 \\ \hline \end{array}$	$\begin{array}{r} 3 \\ +5 \\ \hline \end{array}$	$\begin{array}{r} 3 \\ +6 \\ \hline \end{array}$
$\begin{array}{r} 4 \\ +0 \\ \hline \end{array}$	$\begin{array}{r} 4 \\ +1 \\ \hline \end{array}$	$\begin{array}{r} 4 \\ +2 \\ \hline \end{array}$

Addition Fact Cards

$\begin{array}{r} 4 \\ +3 \\ \hline \end{array}$	$\begin{array}{r} 4 \\ +4 \\ \hline \end{array}$	$\begin{array}{r} 4 \\ +5 \\ \hline \end{array}$
$\begin{array}{r} 5 \\ +0 \\ \hline \end{array}$	$\begin{array}{r} 5 \\ +1 \\ \hline \end{array}$	$\begin{array}{r} 5 \\ +2 \\ \hline \end{array}$
$\begin{array}{r} 5 \\ +3 \\ \hline \end{array}$	$\begin{array}{r} 5 \\ +4 \\ \hline \end{array}$	$\begin{array}{r} 6 \\ +0 \\ \hline \end{array}$

$\begin{array}{r} 6 \\ +1 \\ \hline \end{array}$	$\begin{array}{r} 6 \\ +2 \\ \hline \end{array}$	$\begin{array}{r} 6 \\ +3 \\ \hline \end{array}$
$\begin{array}{r} 7 \\ +0 \\ \hline \end{array}$	$\begin{array}{r} 7 \\ +1 \\ \hline \end{array}$	$\begin{array}{r} 7 \\ +2 \\ \hline \end{array}$
$\begin{array}{r} 8 \\ +0 \\ \hline \end{array}$	$\begin{array}{r} 8 \\ +1 \\ \hline \end{array}$	$\begin{array}{r} 9 \\ +0 \\ \hline \end{array}$

Addition Fact Cards

$$\begin{array}{r} 1 \\ +9 \\ \hline \end{array}$$

$$\begin{array}{r} 2 \\ +8 \\ \hline \end{array}$$

$$\begin{array}{r} 2 \\ +9 \\ \hline \end{array}$$

$$\begin{array}{r} 3 \\ +7 \\ \hline \end{array}$$

$$\begin{array}{r} 3 \\ +8 \\ \hline \end{array}$$

$$\begin{array}{r} 3 \\ +9 \\ \hline \end{array}$$

$$\begin{array}{r} 4 \\ +6 \\ \hline \end{array}$$

$$\begin{array}{r} 4 \\ +7 \\ \hline \end{array}$$

$$\begin{array}{r} 4 \\ +8 \\ \hline \end{array}$$

Addition Fact Cards

4	5	5
$+9$	$+5$	$+6$
5	5	5
$+7$	$+8$	$+9$
6	6	6
$+4$	$+5$	$+6$

Addition Fact Cards

$$6$$
$$+7$$

$$6$$
$$+8$$

$$6$$
$$+9$$

$$7$$
$$+3$$

$$7$$
$$+4$$

$$7$$
$$+5$$

$$7$$
$$+6$$

$$7$$
$$+7$$

$$7$$
$$+8$$

$\begin{array}{r} 7 \\ +\ 9 \\ \hline \end{array}$	$\begin{array}{r} 8 \\ +\ 2 \\ \hline \end{array}$	$\begin{array}{r} 8 \\ +\ 3 \\ \hline \end{array}$
$\begin{array}{r} 8 \\ +\ 4 \\ \hline \end{array}$	$\begin{array}{r} 8 \\ +\ 5 \\ \hline \end{array}$	$\begin{array}{r} 8 \\ +\ 6 \\ \hline \end{array}$
$\begin{array}{r} 8 \\ +\ 7 \\ \hline \end{array}$	$\begin{array}{r} 8 \\ +\ 8 \\ \hline \end{array}$	$\begin{array}{r} 8 \\ +\ 9 \\ \hline \end{array}$

Addition Fact Cards

9 +1	9 +2	9 +3
9 +4	9 +5	9 +6
9 +7	9 +8	9 +9

Addition Fact Cards

R12

Addition Table

Base-Ten Materials (tens, ones)

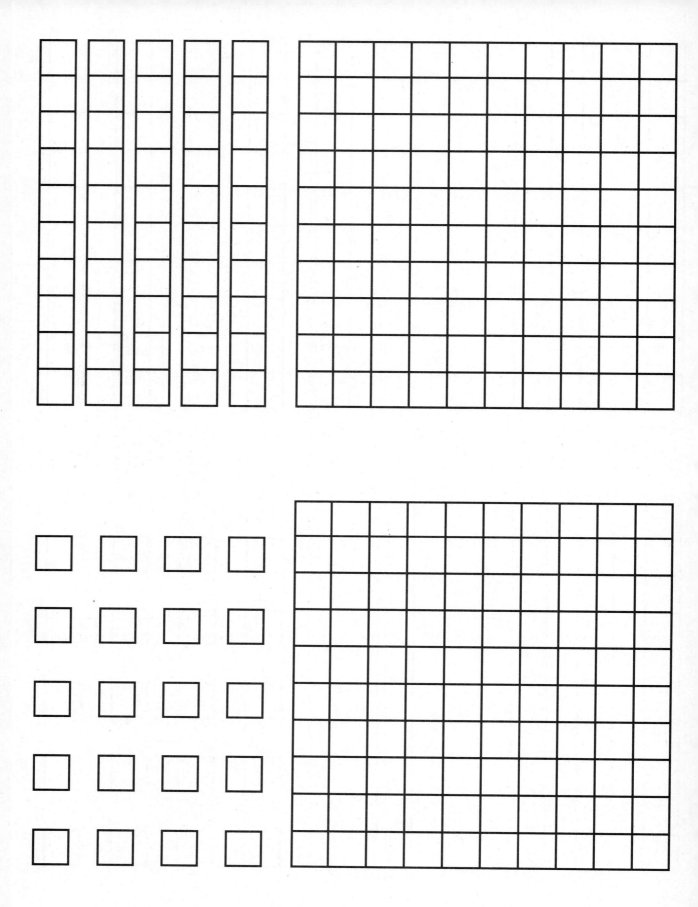

Base-Ten Materials (hundreds, tens, ones)

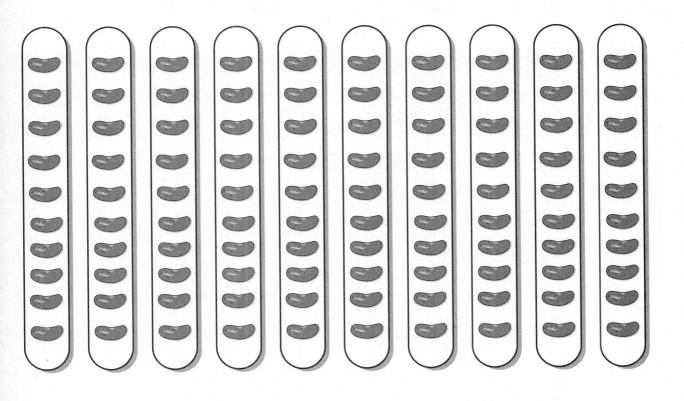

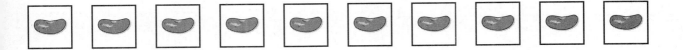

Bean Sticks and Beans

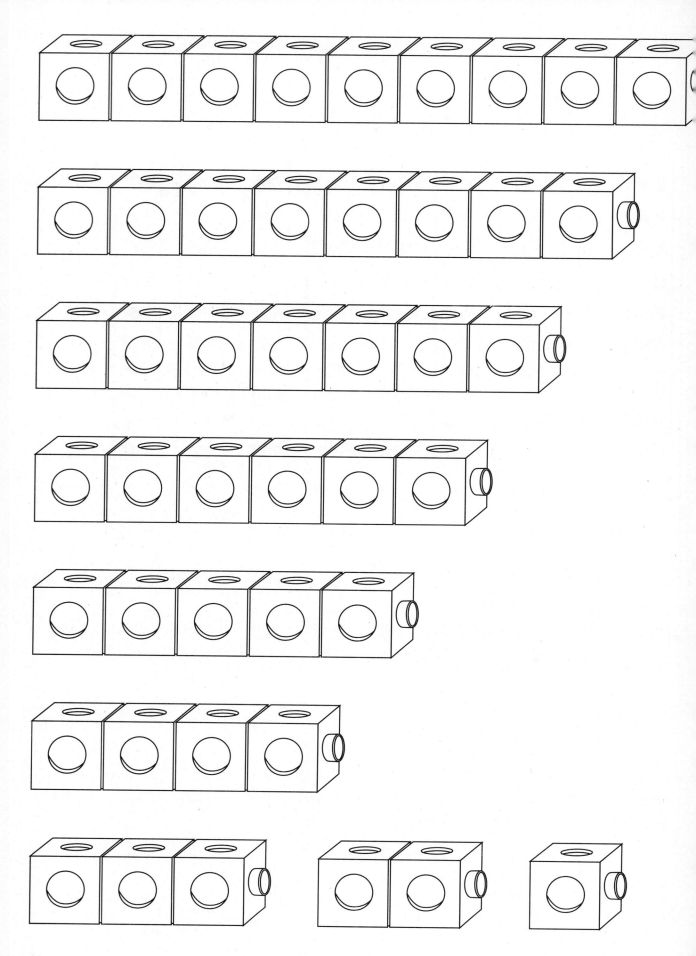

Connecting Cube Patterns (one to nine)

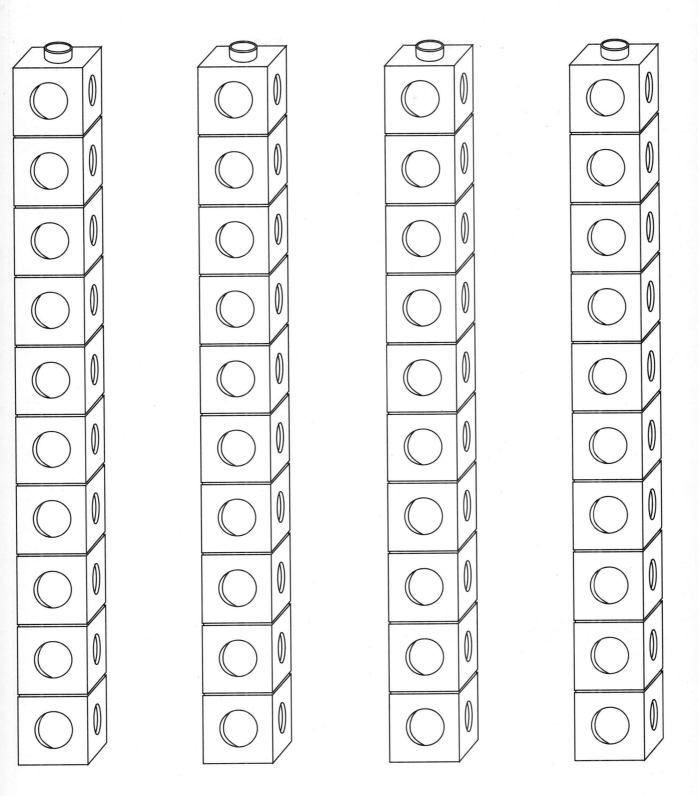

Connecting Cube Patterns (tens)

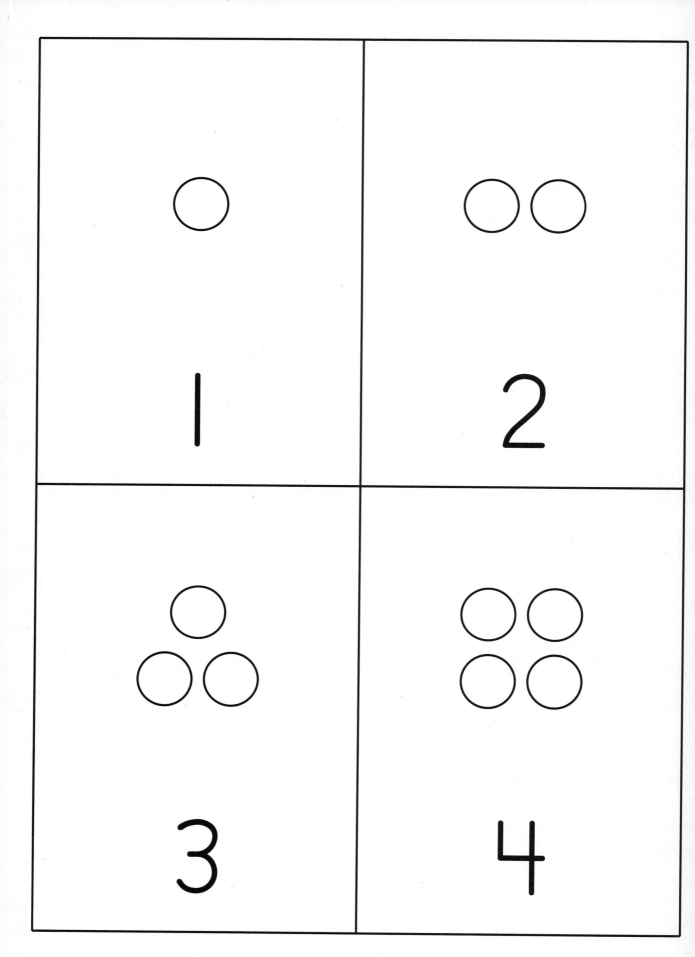

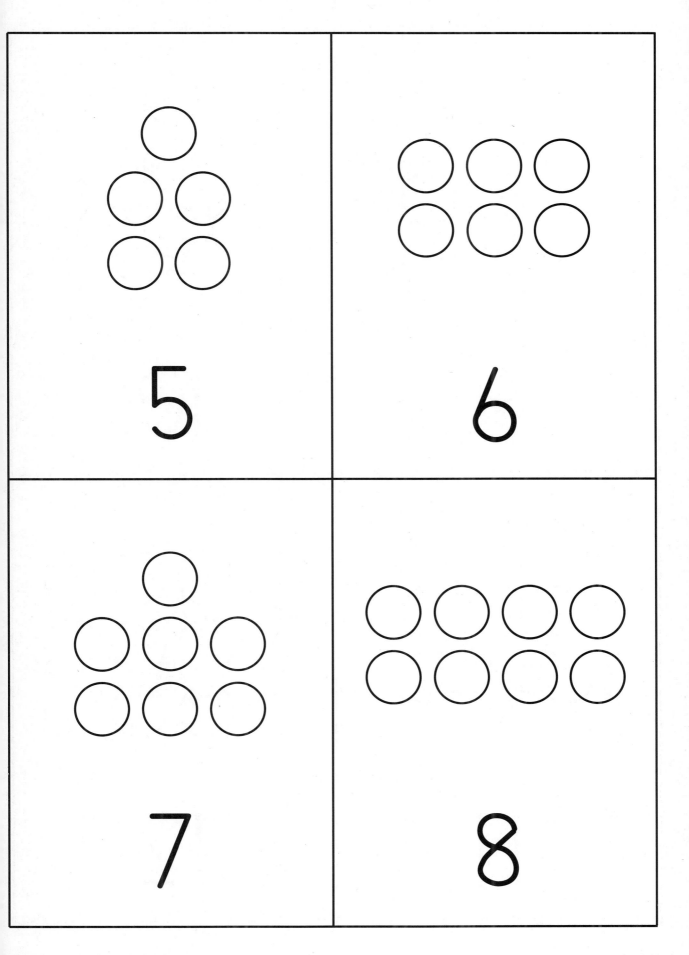

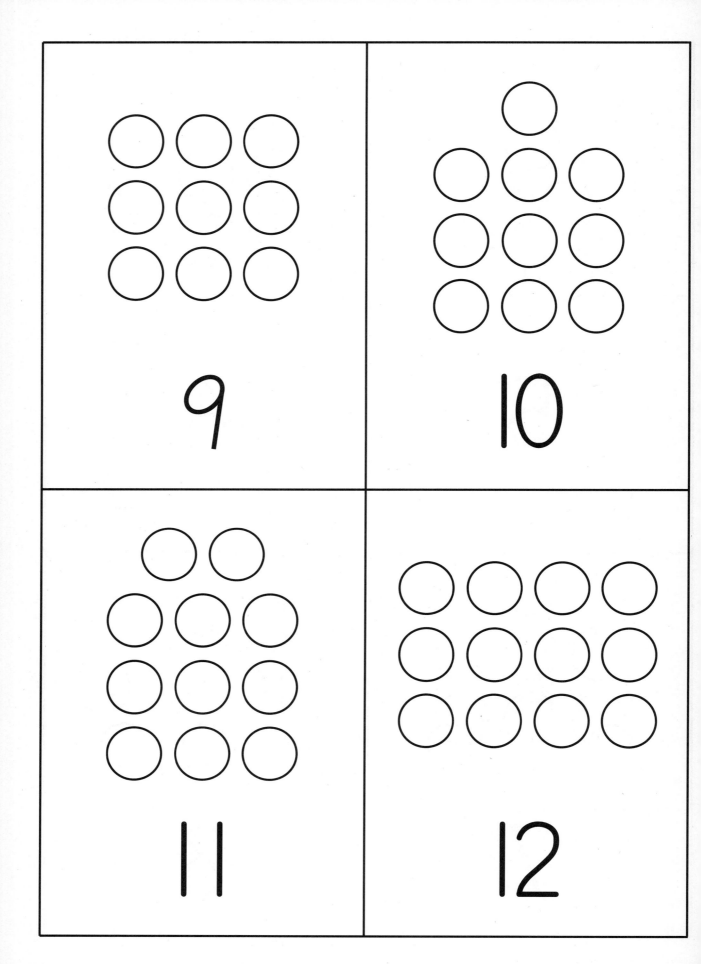

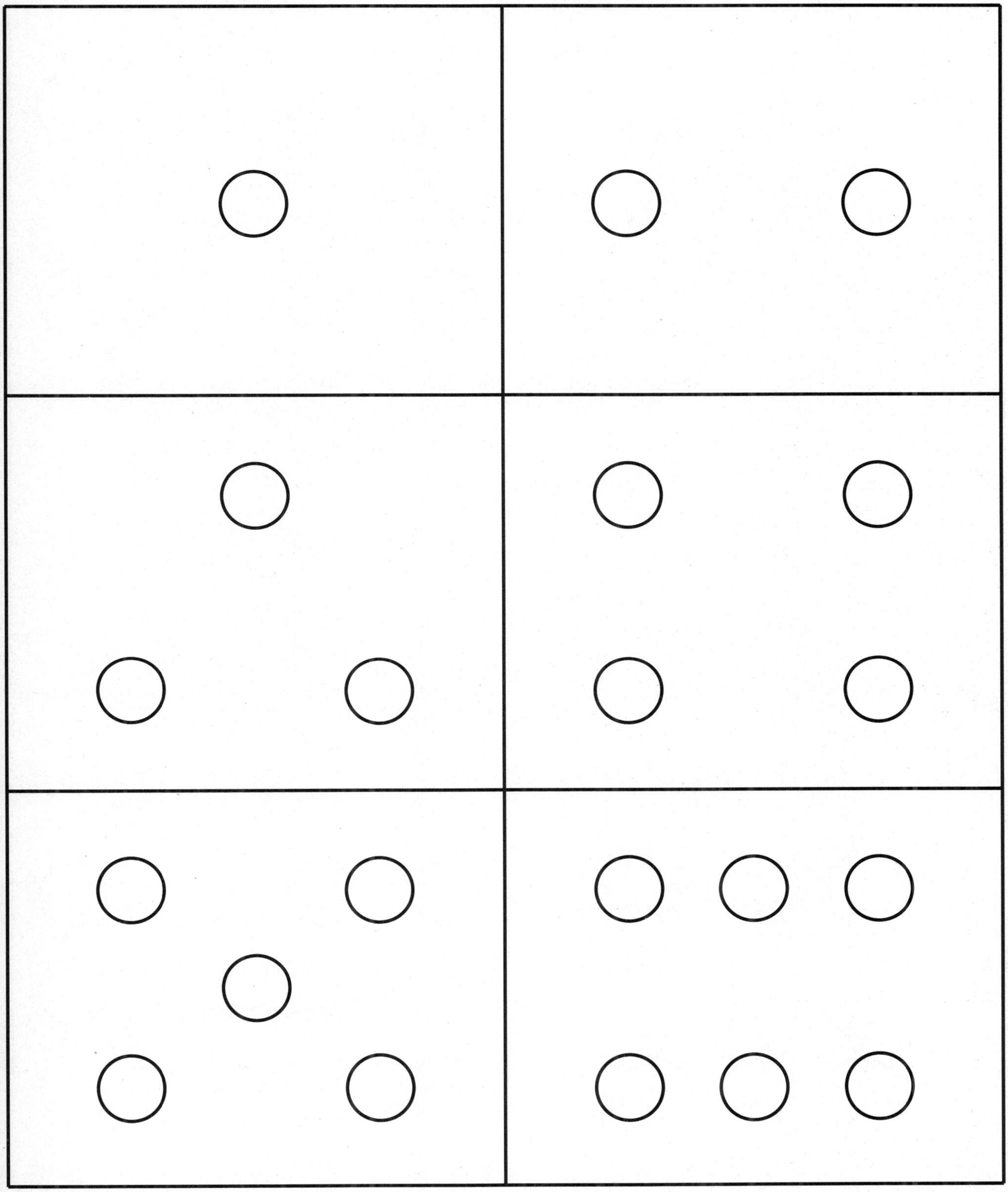

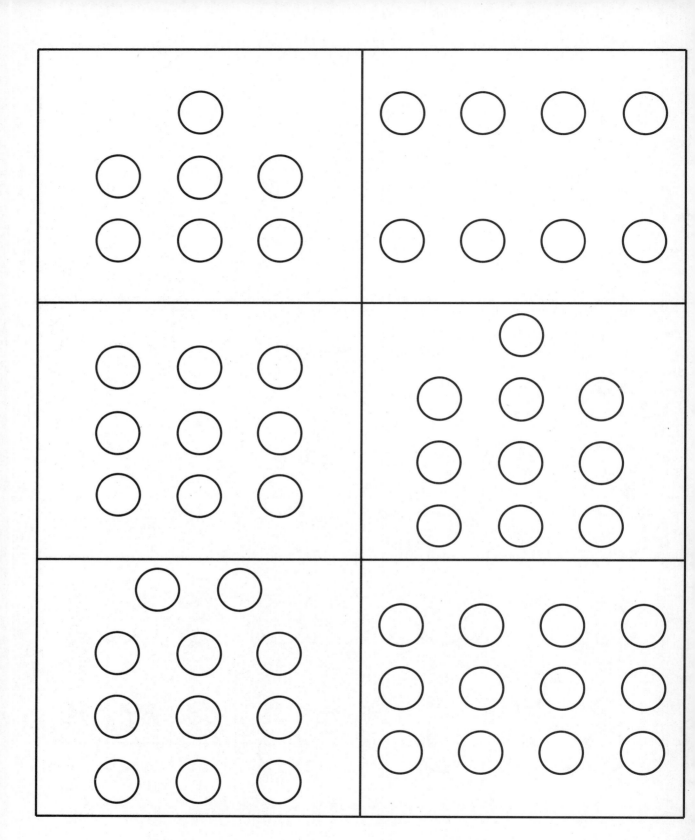

1	2	3	4	5	6	7	8	9	10
11	12	13	14	15	16	17	18	19	20
21	22	23	24	25	26	27	28	29	30
31	32	33	34	35	36	37	38	39	40
41	42	43	44	45	46	47	48	49	50
51	52	53	54	55	56	57	58	59	60
61	62	63	64	65	66	67	68	69	70
71	72	73	74	75	76	77	78	79	80
81	82	83	84	85	86	87	88	89	90
91	92	93	94	95	96	97	98	99	100

Hundred Chart

zero	one
two	three
four	five

six	seven
eight	nine
ten	twenty

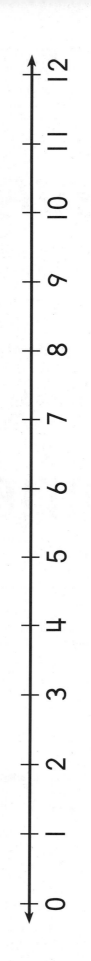

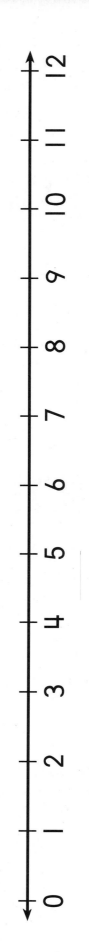

Number Lines (0–12)

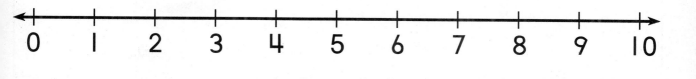

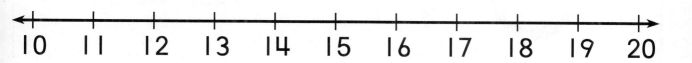

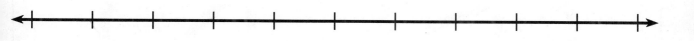

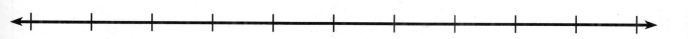

Number Lines (0–10, 10–20, blank)

4	0
5	1
6	2
7	3

12	8
13	9
14	10
15	11

20	16
21	17
22	18
23	19

24	28
25	29
26	30
27	31

32	36
33	37
34	38
35	39

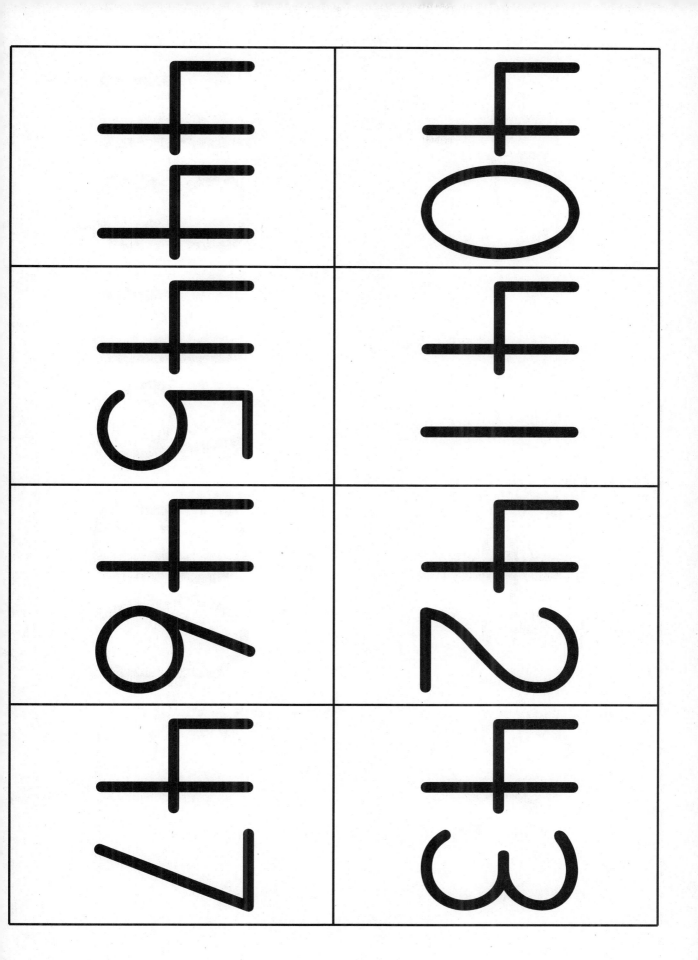

	48
+	
−	49
=	50
÷	

0 0

1 1

2 2

3 3

4 4

5 5

6 6

7 7

8 8

9 9

10 10

first	second
third	fourth
fifth	sixth

Ordinal Number Word Cards (first through sixth)

seventh	eighth
ninth	tenth
eleventh	twelfth

tens	ones

tens	ones

tens	ones

tens	ones

tens	ones

tens	ones

tens	ones

tens	ones

tens	ones

tens	ones

tens	ones

tens	ones

Place-Value Charts (tens, ones)

hundreds	tens	ones

hundreds	tens	ones

hundreds	tens	ones

hundreds	tens	ones

hundreds	tens	ones

hundreds	tens	ones

hundreds	tens	ones

hundreds	tens	ones

Place-Value Charts (hundreds, tens, ones)

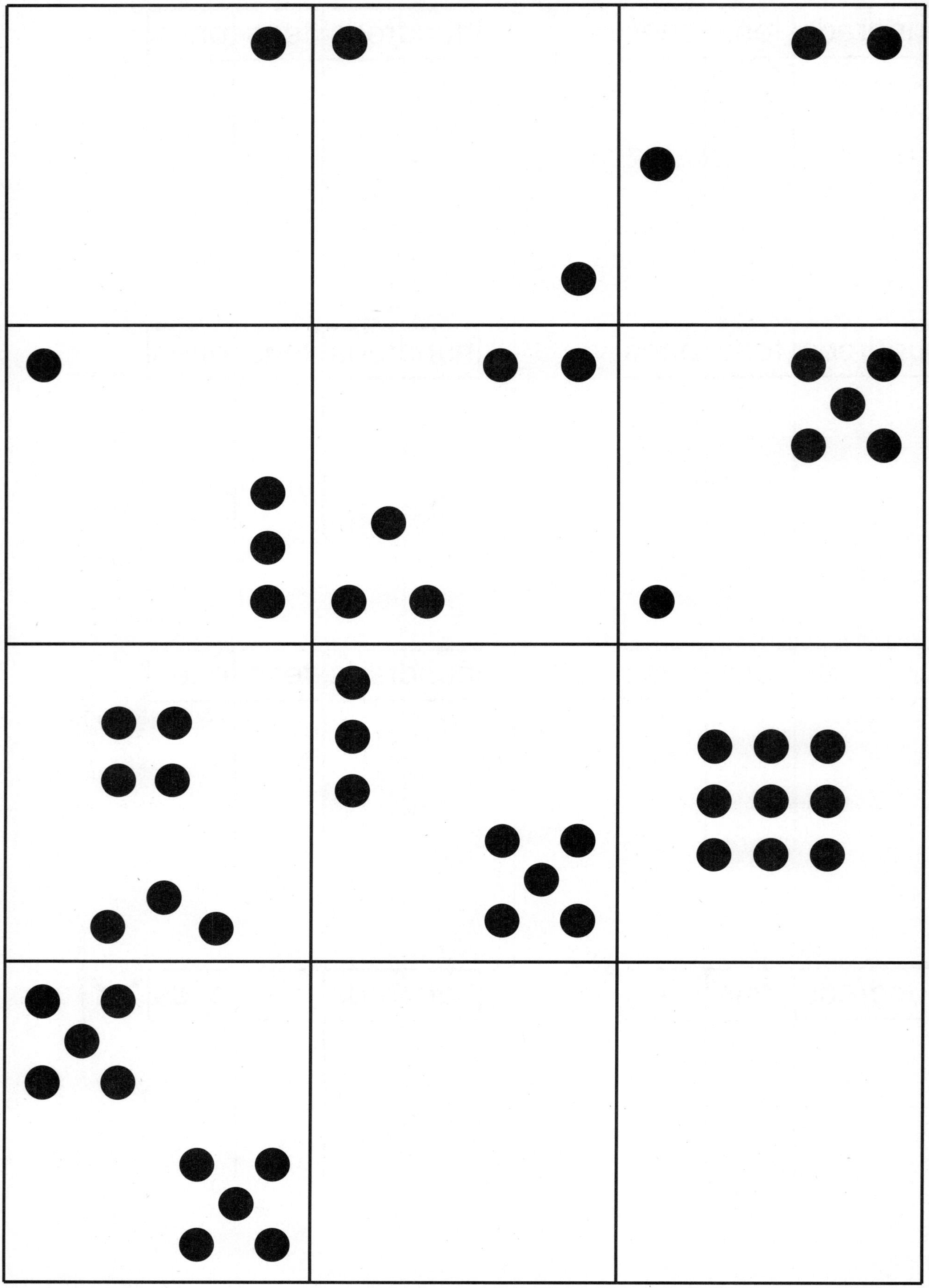

9 −9 ———	9 −8 ———	9 −7 ———
9 −6 ———	9 −5 ———	9 −4 ———
9 −3 ———	9 −2 ———	9 −1 ———

$$\begin{array}{r} 9 \\ -\ 0 \\ \hline \end{array}$$

$$\begin{array}{r} 8 \\ -\ 8 \\ \hline \end{array}$$

$$\begin{array}{r} 8 \\ -\ 7 \\ \hline \end{array}$$

$$\begin{array}{r} 8 \\ -\ 6 \\ \hline \end{array}$$

$$\begin{array}{r} 8 \\ -\ 5 \\ \hline \end{array}$$

$$\begin{array}{r} 8 \\ -\ 4 \\ \hline \end{array}$$

$$\begin{array}{r} 8 \\ -\ 3 \\ \hline \end{array}$$

$$\begin{array}{r} 8 \\ -\ 2 \\ \hline \end{array}$$

$$\begin{array}{r} 8 \\ -\ 1 \\ \hline \end{array}$$

$$\begin{array}{r} 8 \\ -\ 0 \\ \hline \end{array}$$

$$\begin{array}{r} 7 \\ -\ 7 \\ \hline \end{array}$$

$$\begin{array}{r} 7 \\ -\ 6 \\ \hline \end{array}$$

$$\begin{array}{r} 7 \\ -\ 5 \\ \hline \end{array}$$

$$\begin{array}{r} 7 \\ -\ 4 \\ \hline \end{array}$$

$$\begin{array}{r} 7 \\ -\ 3 \\ \hline \end{array}$$

$$\begin{array}{r} 7 \\ -\ 2 \\ \hline \end{array}$$

$$\begin{array}{r} 7 \\ -\ 1 \\ \hline \end{array}$$

$$\begin{array}{r} 7 \\ -\ 0 \\ \hline \end{array}$$

6 − 6 ————	6 − 5 ————	6 − 4 ————
6 − 3 ————	6 − 2 ————	6 − 1 ————
6 − 0 ————	5 − 5 ————	5 − 4 ————

Subtraction Fact Cards

$\begin{array}{r} 5 \\ -3 \\ \hline \end{array}$	$\begin{array}{r} 5 \\ -2 \\ \hline \end{array}$	$\begin{array}{r} 5 \\ -1 \\ \hline \end{array}$
$\begin{array}{r} 5 \\ -0 \\ \hline \end{array}$	$\begin{array}{r} 4 \\ -4 \\ \hline \end{array}$	$\begin{array}{r} 4 \\ -3 \\ \hline \end{array}$
$\begin{array}{r} 4 \\ -2 \\ \hline \end{array}$	$\begin{array}{r} 4 \\ -1 \\ \hline \end{array}$	$\begin{array}{r} 4 \\ -0 \\ \hline \end{array}$

$\begin{array}{r} 3 \\ -\ 3 \\ \hline \end{array}$	$\begin{array}{r} 3 \\ -\ 2 \\ \hline \end{array}$	$\begin{array}{r} 3 \\ -\ 1 \\ \hline \end{array}$
$\begin{array}{r} 3 \\ -\ 0 \\ \hline \end{array}$	$\begin{array}{r} 2 \\ -\ 2 \\ \hline \end{array}$	$\begin{array}{r} 2 \\ -\ 1 \\ \hline \end{array}$
$\begin{array}{r} 2 \\ -\ 0 \\ \hline \end{array}$	$\begin{array}{r} 1 \\ -\ 1 \\ \hline \end{array}$	$\begin{array}{r} 1 \\ -\ 0 \\ \hline \end{array}$

10 − 1 ————	10 − 2 ————	10 − 3 ————
10 − 4 ————	10 − 5 ————	10 − 6 ————
10 − 7 ————	10 − 8 ————	10 − 9 ————

11 − 2	11 − 3	11 − 4
11 − 5	11 − 6	11 − 7
11 − 8	11 − 9	12 − 3

Subtraction Fact Cards

12	12	12
$-\ 4$	$-\ 5$	$-\ 6$

12	12	12
$-\ 7$	$-\ 8$	$-\ 9$

13	13	13
$-\ 4$	$-\ 5$	$-\ 6$

$\begin{array}{r} 13 \\ -7 \\ \hline \end{array}$	$\begin{array}{r} 13 \\ -8 \\ \hline \end{array}$	$\begin{array}{r} 13 \\ -9 \\ \hline \end{array}$
$\begin{array}{r} 14 \\ -5 \\ \hline \end{array}$	$\begin{array}{r} 14 \\ -6 \\ \hline \end{array}$	$\begin{array}{r} 14 \\ -7 \\ \hline \end{array}$
$\begin{array}{r} 14 \\ -8 \\ \hline \end{array}$	$\begin{array}{r} 14 \\ -9 \\ \hline \end{array}$	$\begin{array}{r} 15 \\ -6 \\ \hline \end{array}$

Subtraction Fact Cards

15 −7	15 −8	15 −9
16 −7	16 −8	16 −9
17 −8	17 −9	18 −9

Subtraction Fact Cards

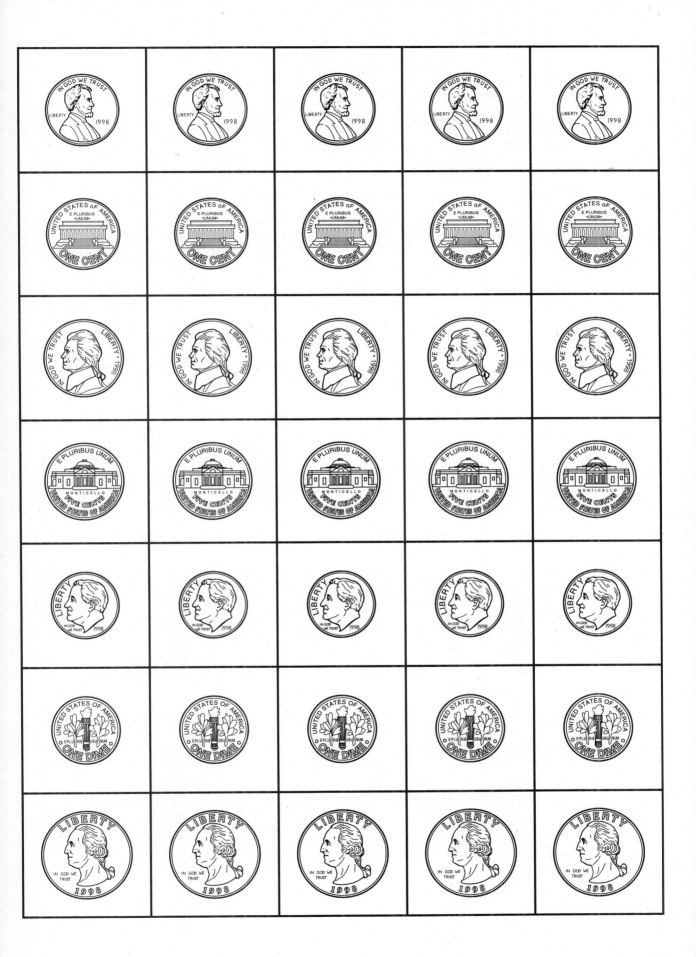

Coins

Coins and Bills

25¢	10¢	5¢	1¢

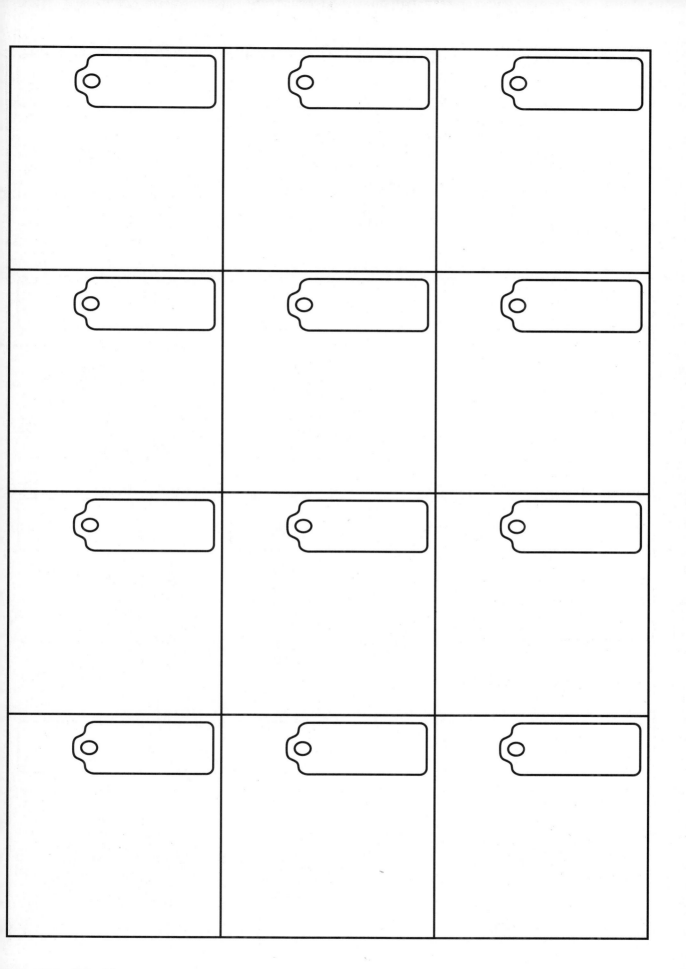

Price Tags (blank)

Toy Pictures (with price tags)

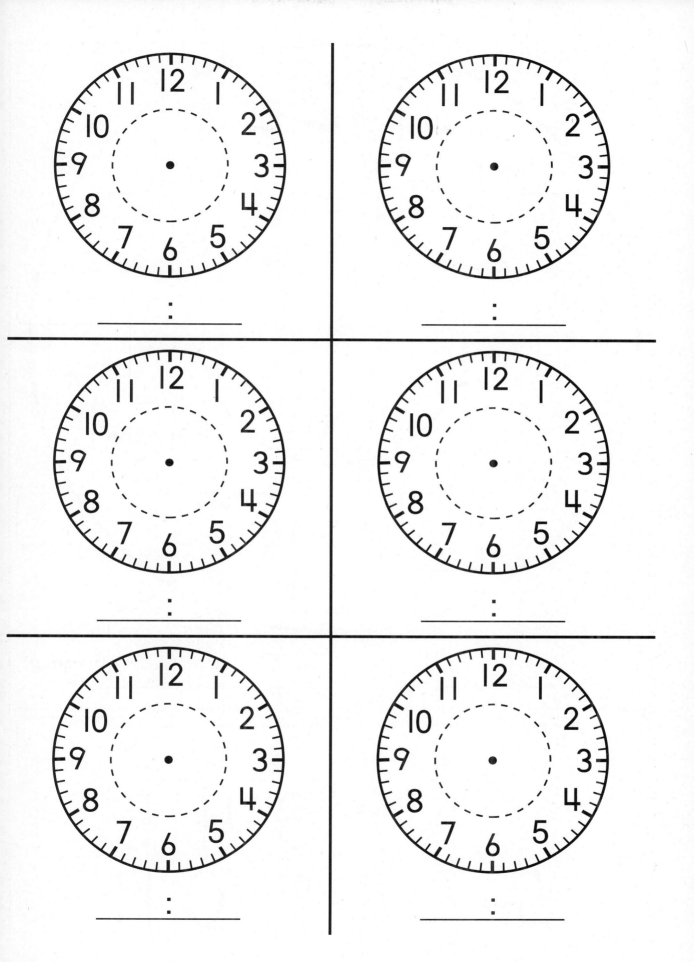

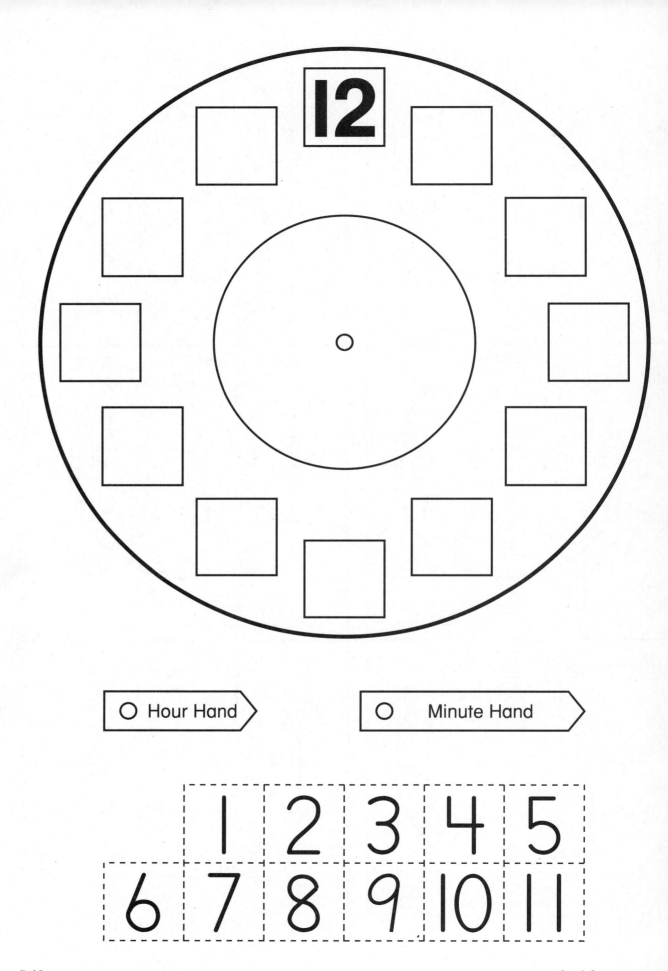

R62

Clockface Pattern

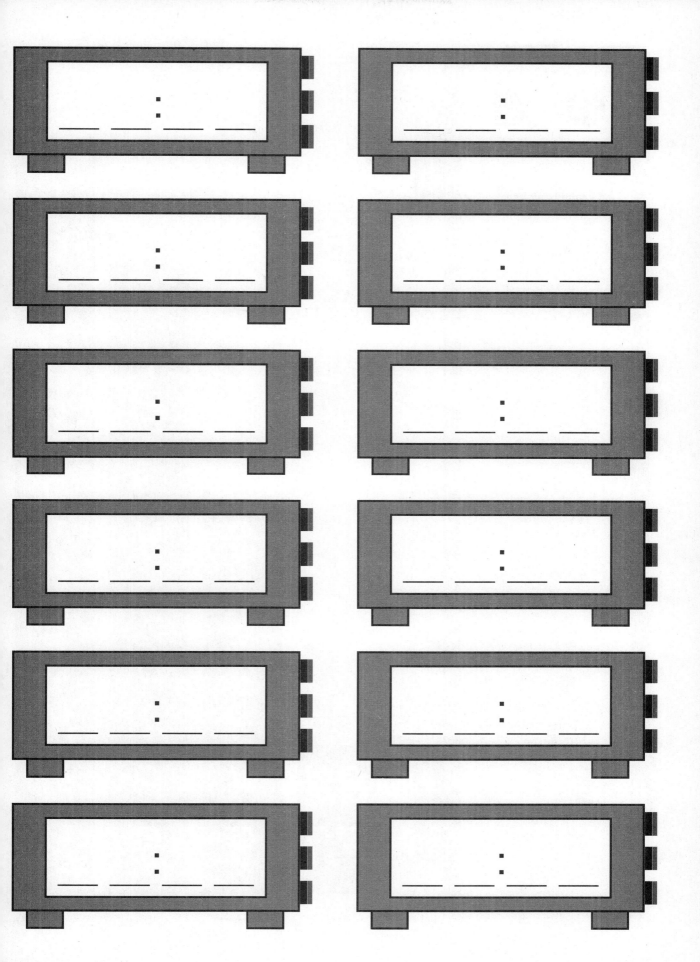

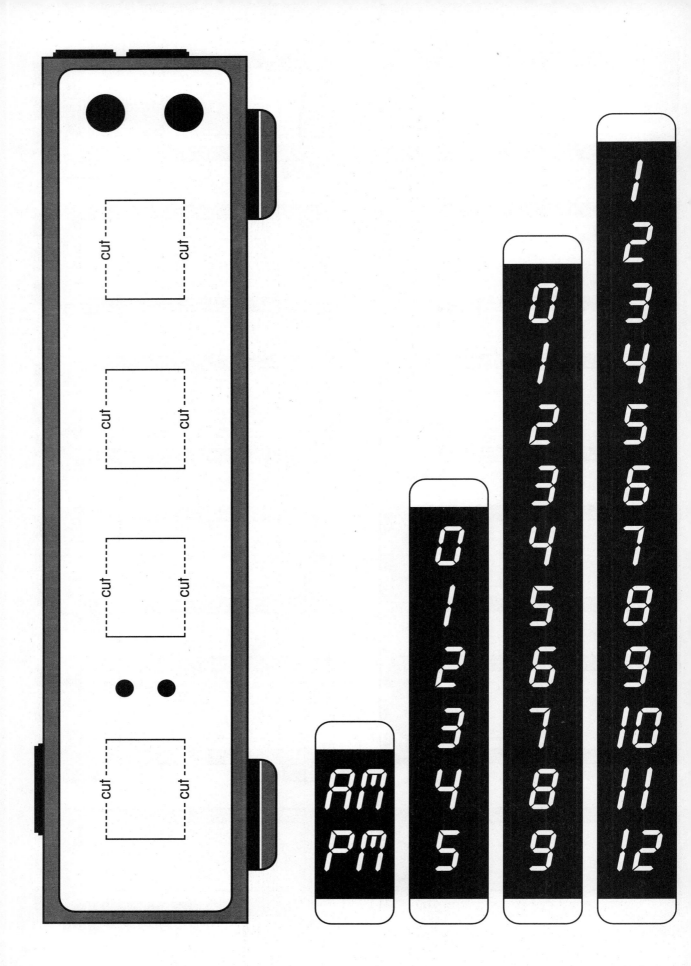

Digital Clock Model

Sunday	Monday	Tuesday	Wednesday	Thursday	Friday	Saturday

Blank Calendar

Name _____

What I Used	Guess	Test

Measurement Recording Sheet

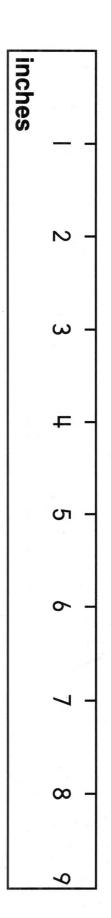

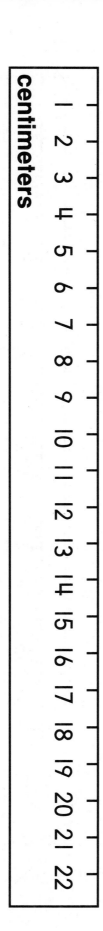

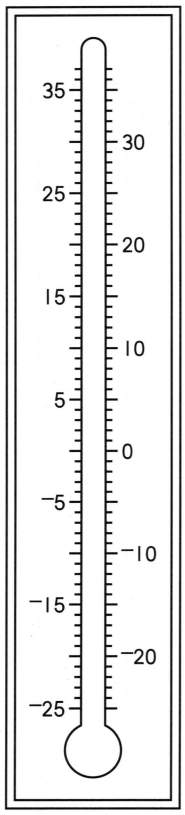

Celsius

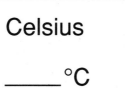

_____ °C

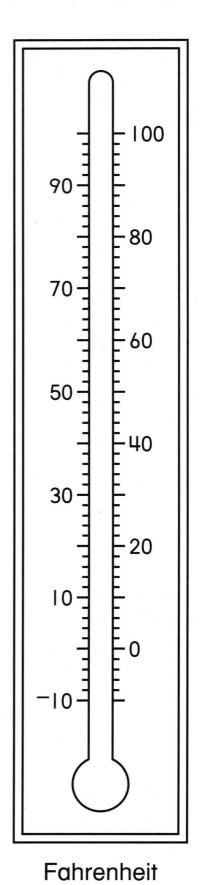

Fahrenheit

_____ °F

Circles

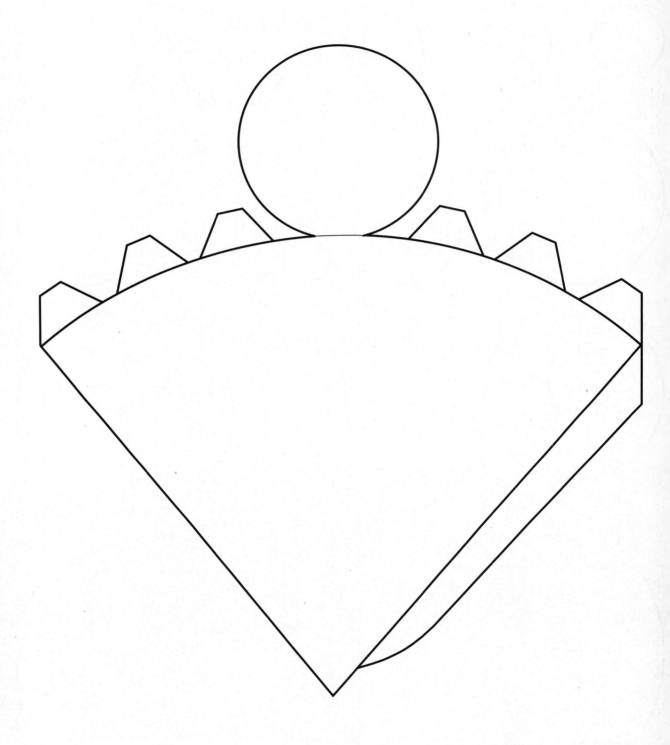

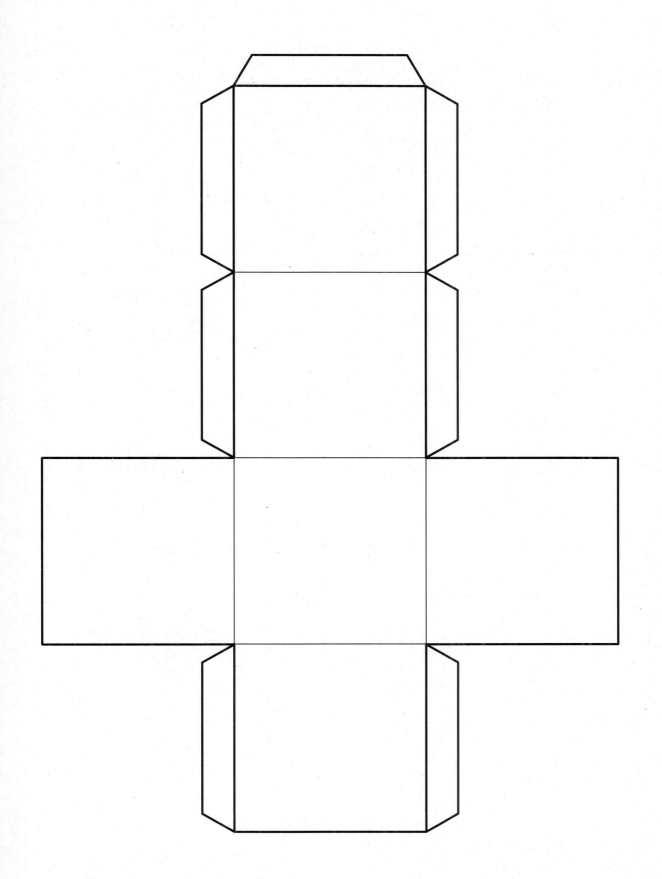

Cube Pattern

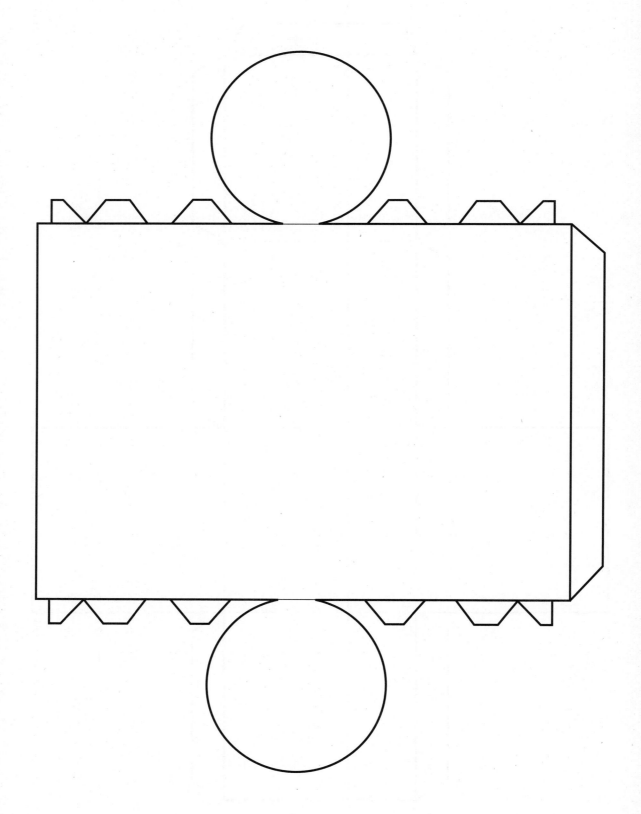

Cylinder Pattern

Dot Paper

R74

Dot Paper (isometric)

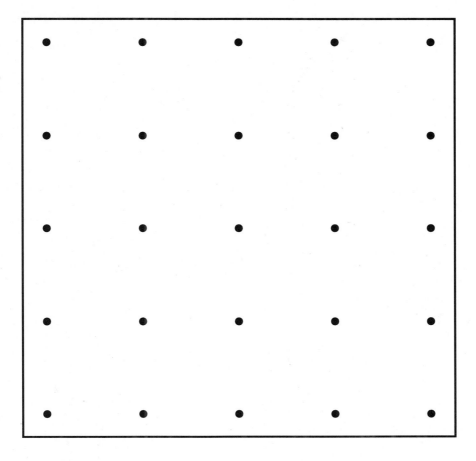

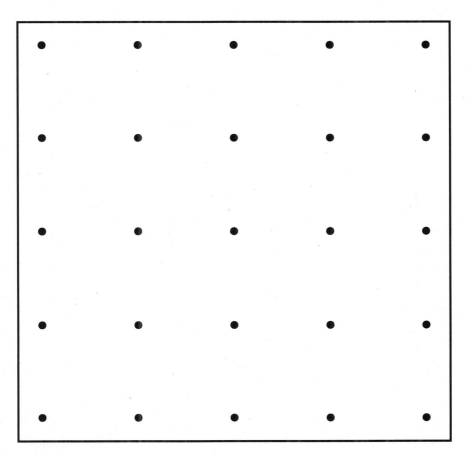

Geoboard Dot Paper

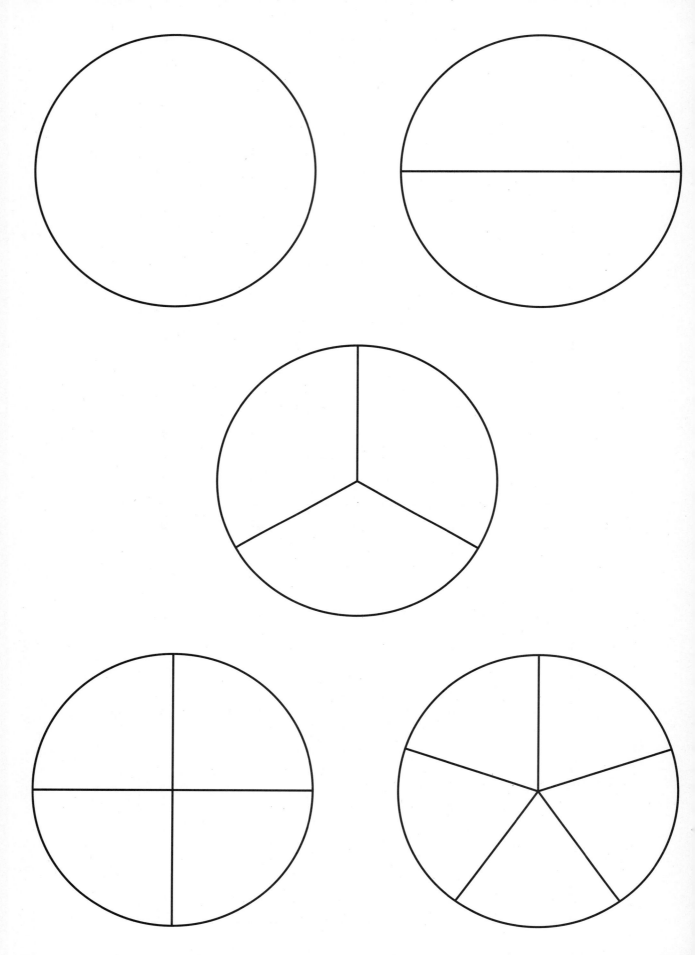

Fraction Circles (whole to fifths)

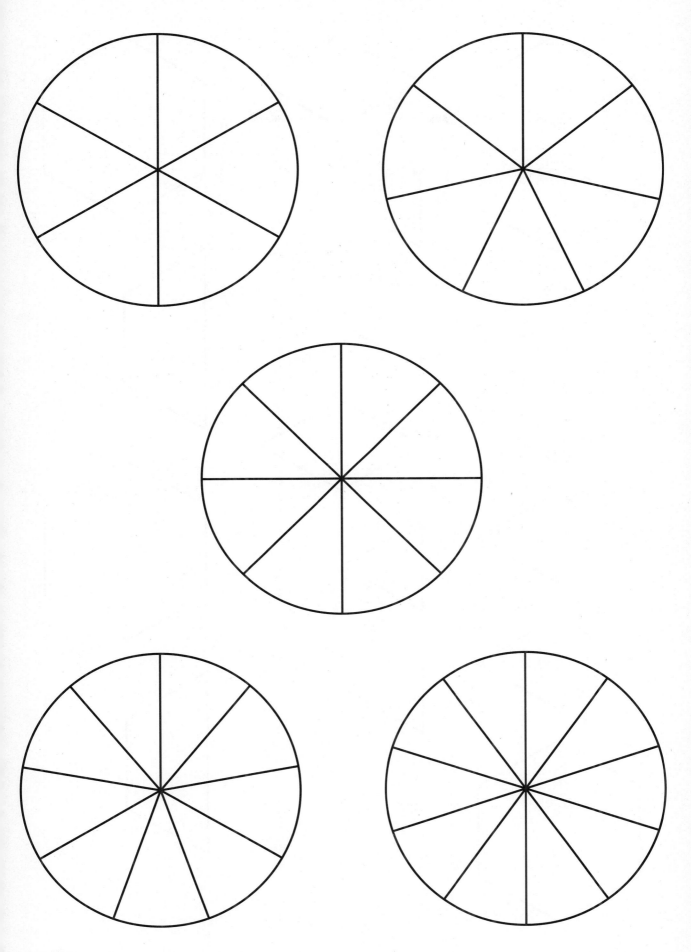

Fraction Circles (sixths to tenths)

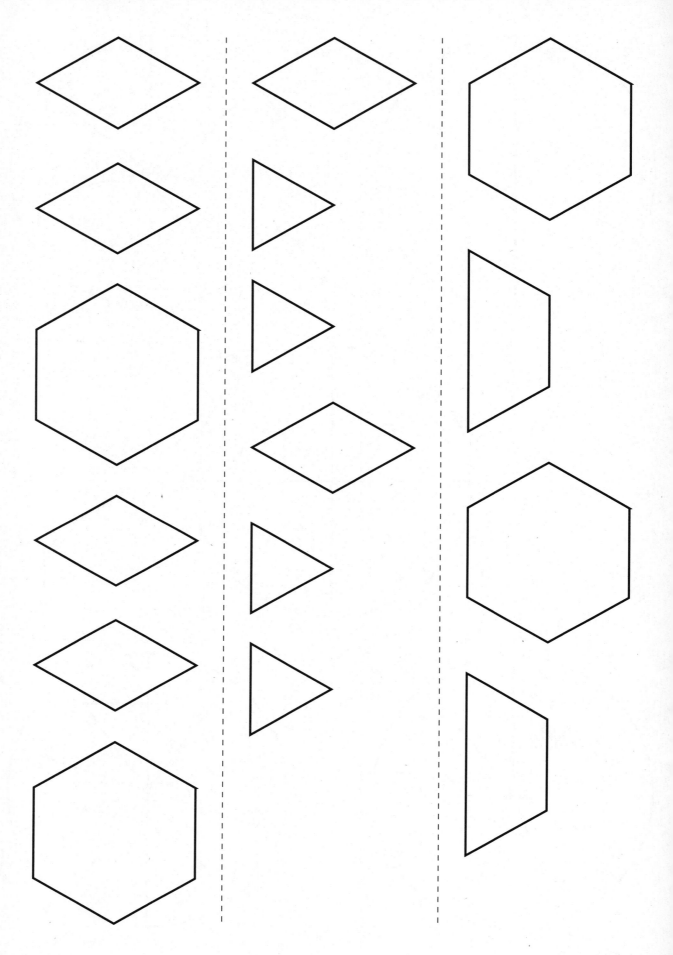

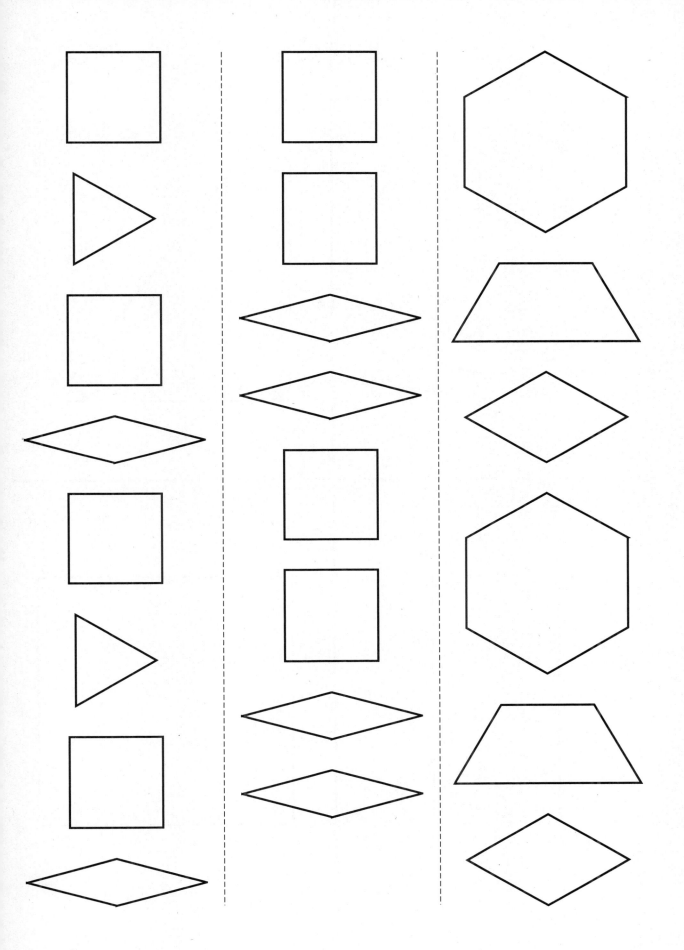

Rectangles

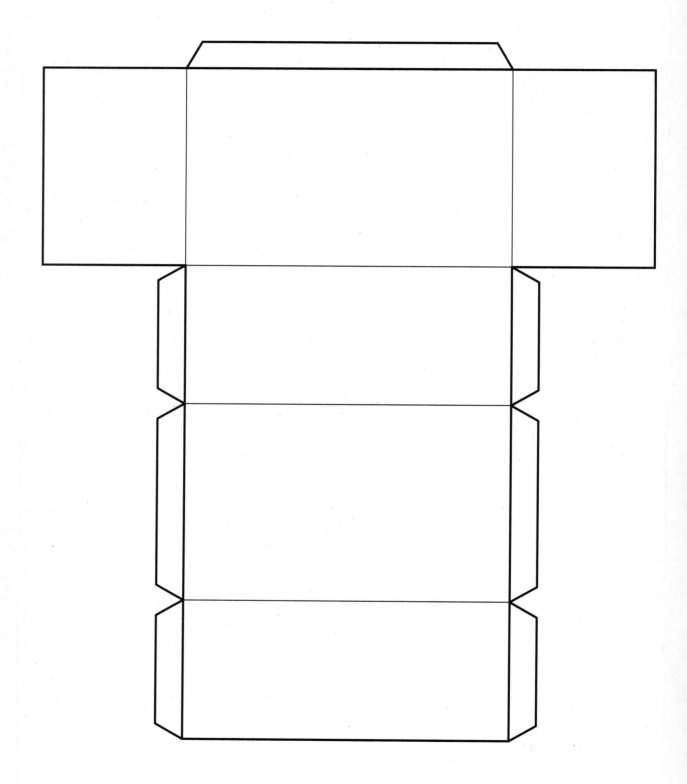

Rectangular Prism Pattern

Squares

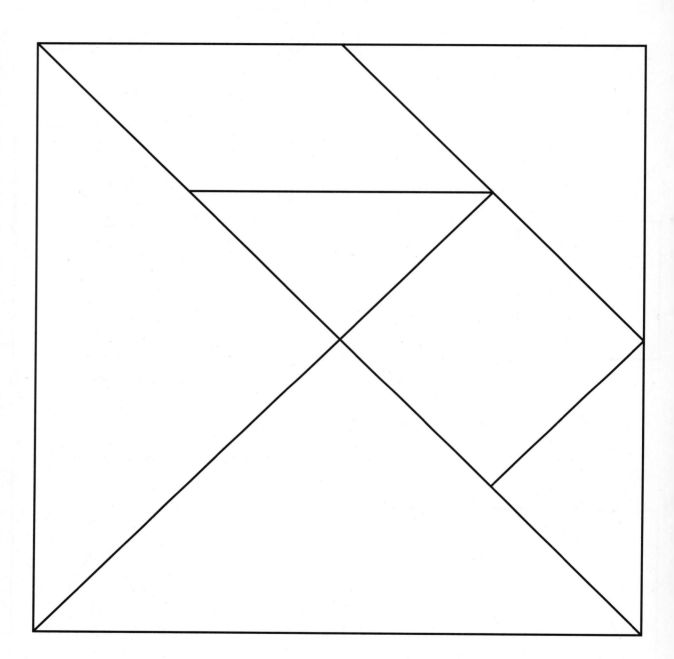

Triangle-Shaped Grid

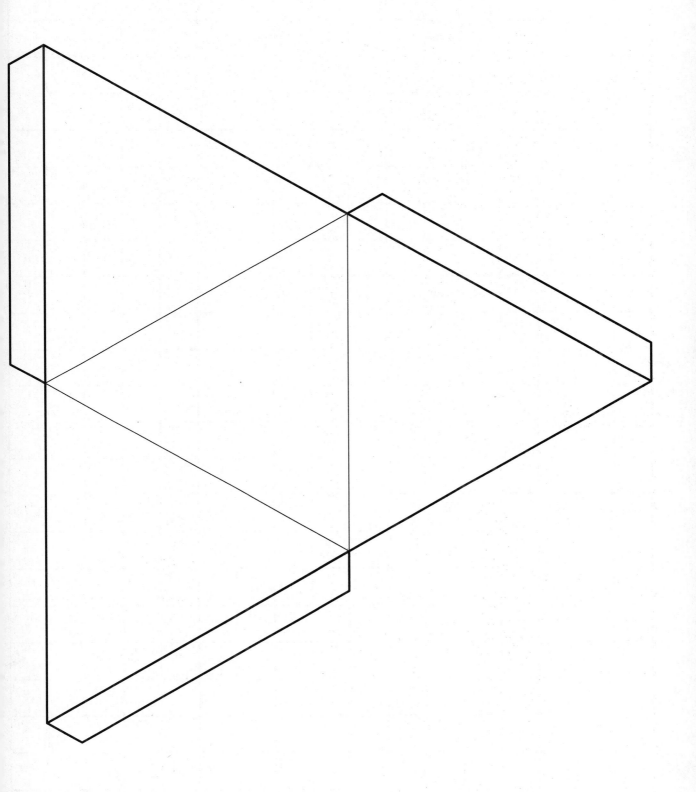

Triangular Pyramid Pattern

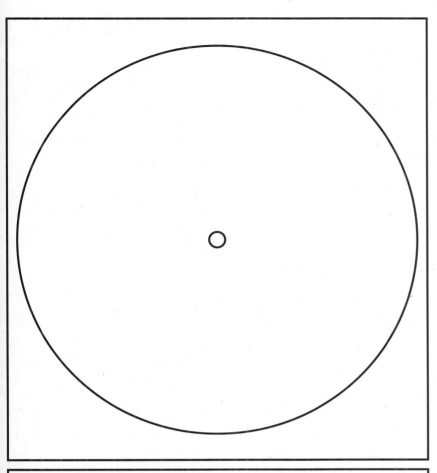

Spinner Tips

How to assemble spinner.
- Glue patterns to oaktag.
- Cut out and attach pointer with a fastener.

Alternative
- Children can use a paper clip and pencil instead.

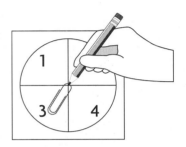

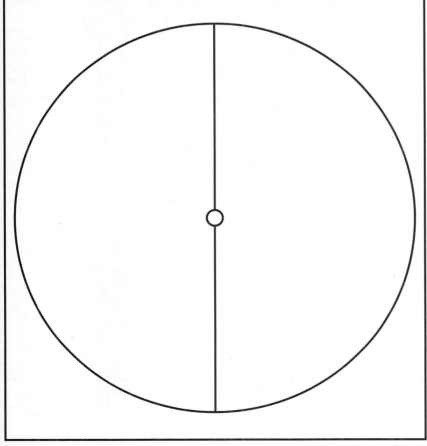

Spinners (blank and 2-section)

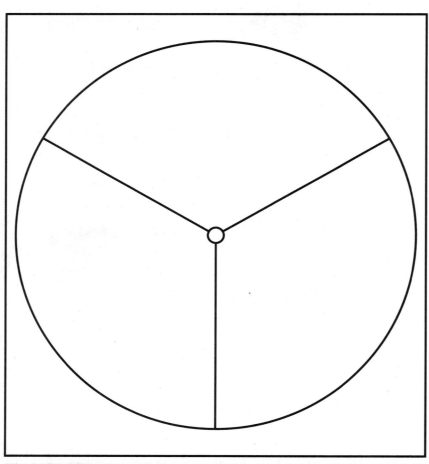

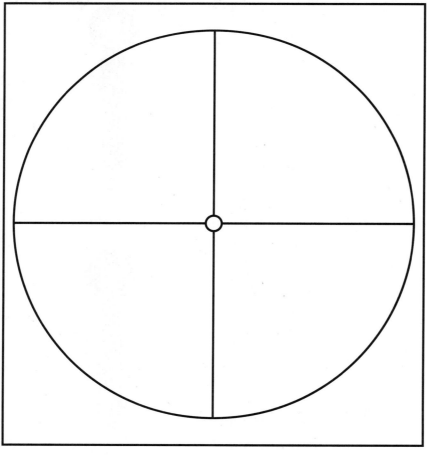

How to assemble spinner.
- Glue patterns to oaktag.
- Cut out and attach pointer with a fastener.

Alternative
- Children can use a paper clip and pencil instead.

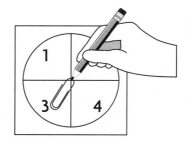

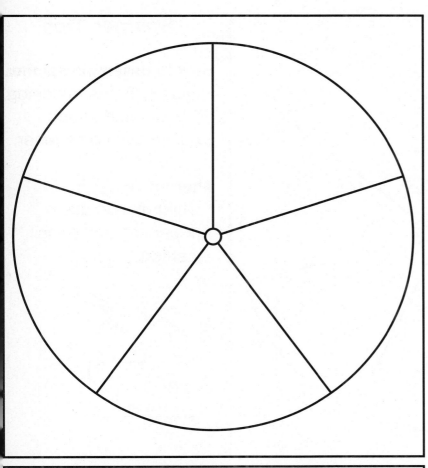

Spinner Tips

How to assemble spinner.
- Glue patterns to oaktag.
- Cut out and attach pointer with a fastener.

Alternative
- Children can use a paper clip and pencil instead.

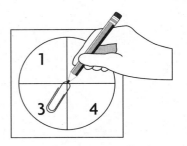

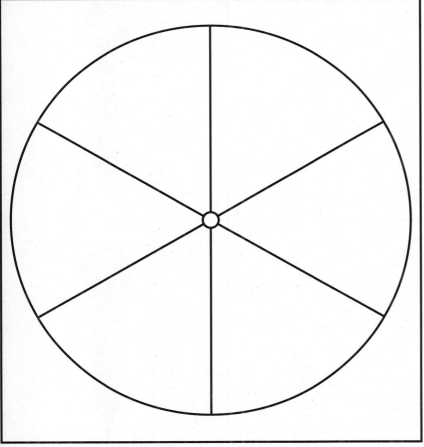

Spinners (5- and 6-section)

R93

Spinner Tips

How to assemble spinner.
- Glue patterns to oaktag.
- Cut out and attach pointer with a fastener.

Alternative
- Children can use a paper clip and pencil instead.

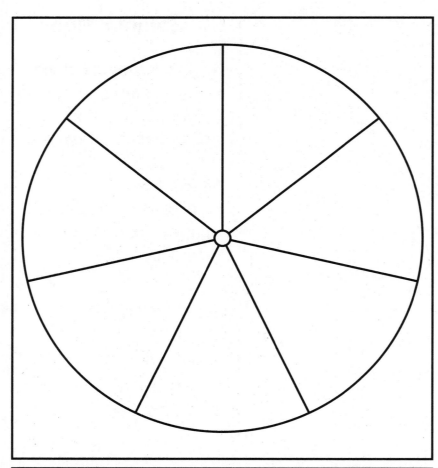

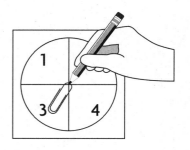

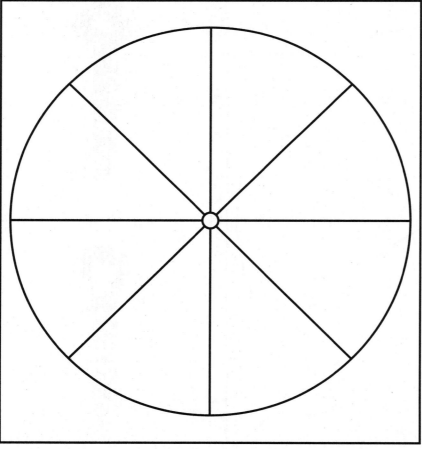

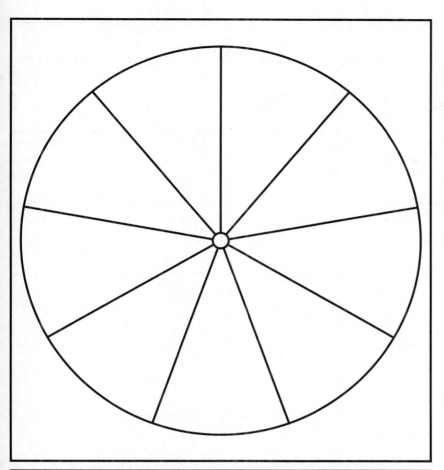

How to assemble spinner.
- Glue patterns to oaktag.
- Cut out and attach pointer with a fastener.

Alternative
- Children can use a paper clip and pencil instead.

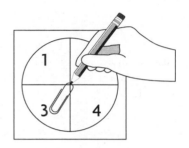

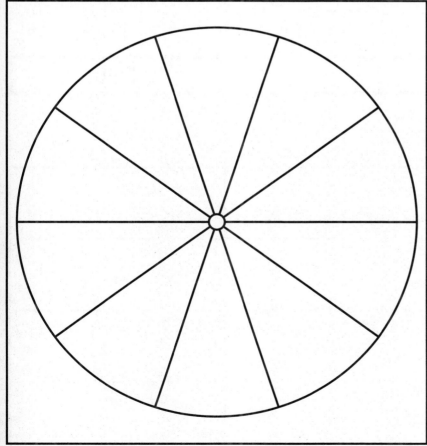

Spinners (9- and 10-section)

		Total

		Total

Tally Table

Finish

Start

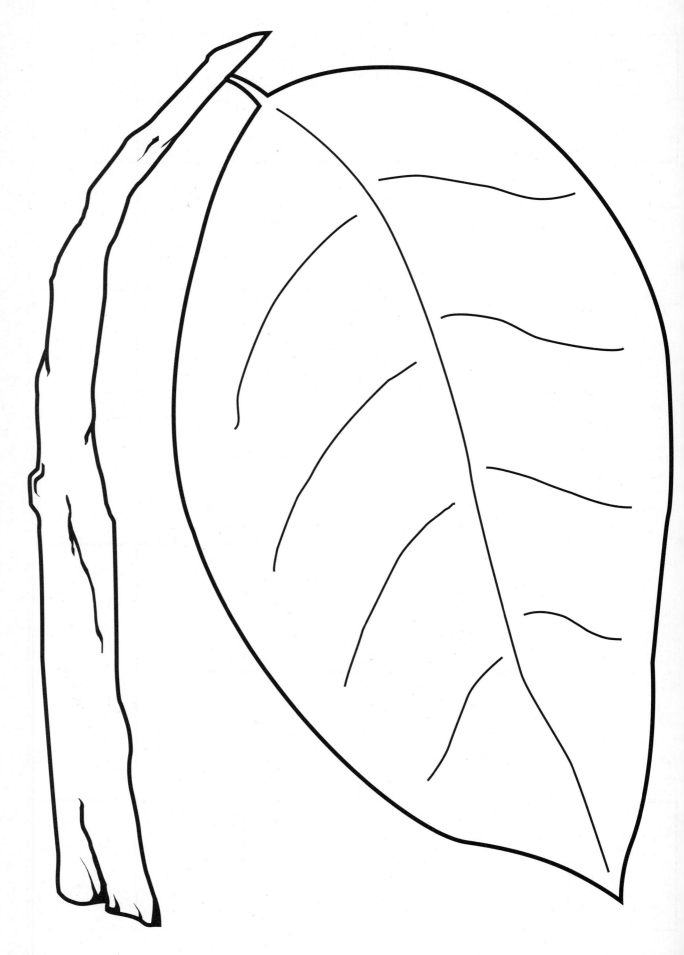

Leaf Workmat

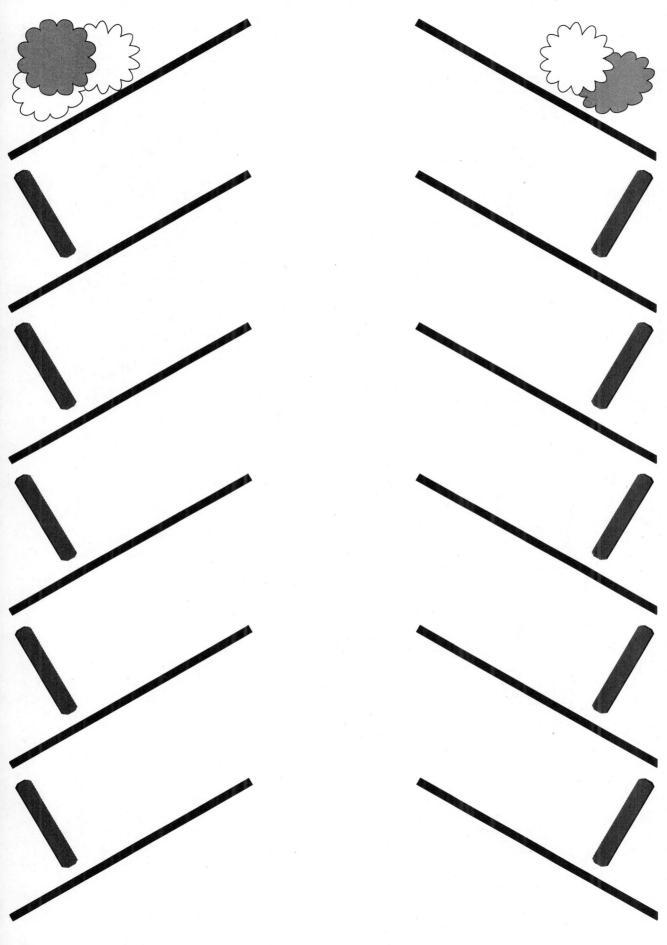

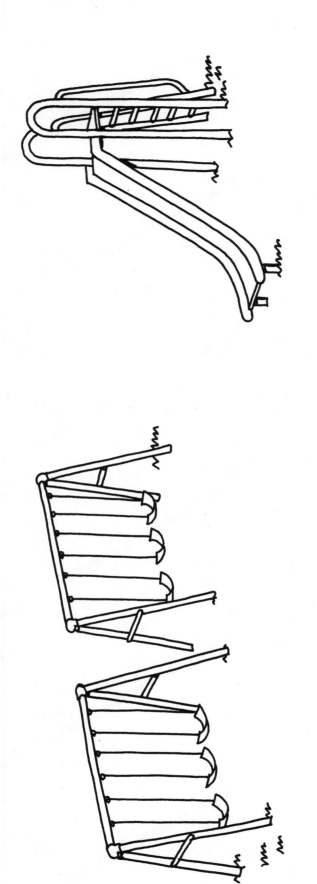
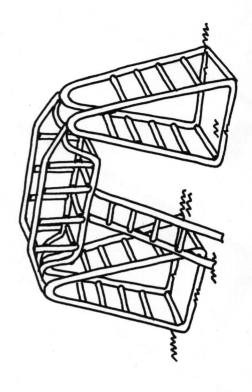

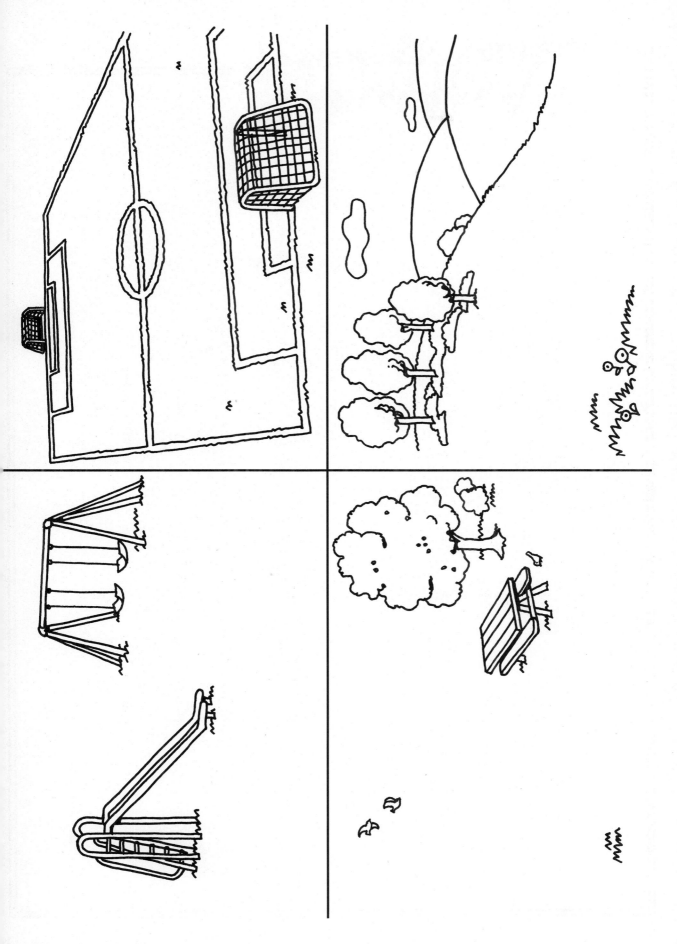

Workmat 1

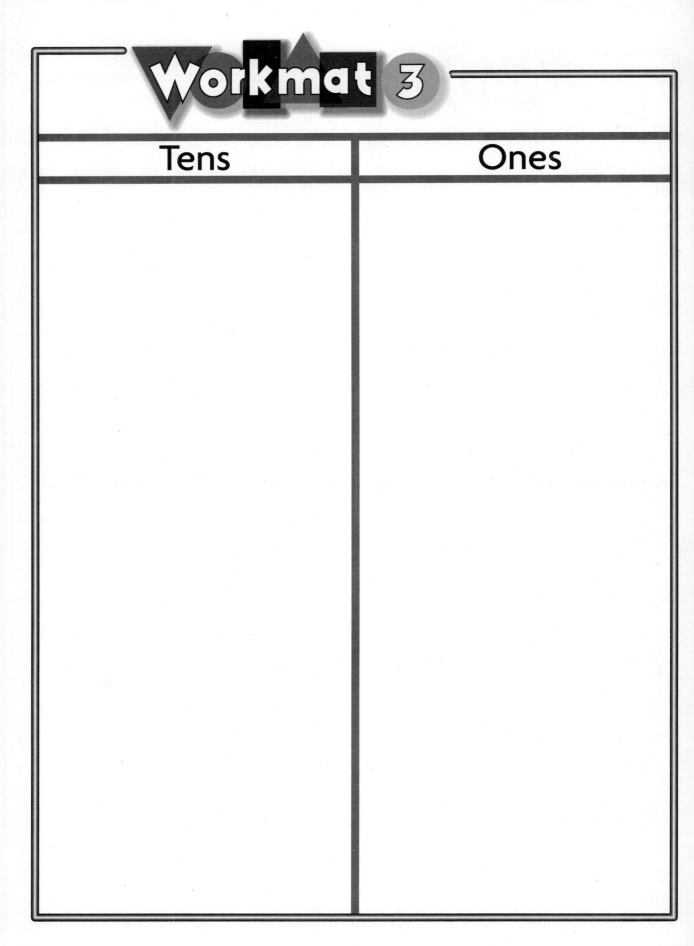

Workmat 3

Tens	Ones

Workmat 4

Penny	Nickel	Dime	Quarter

Hundreds	Tens	Ones

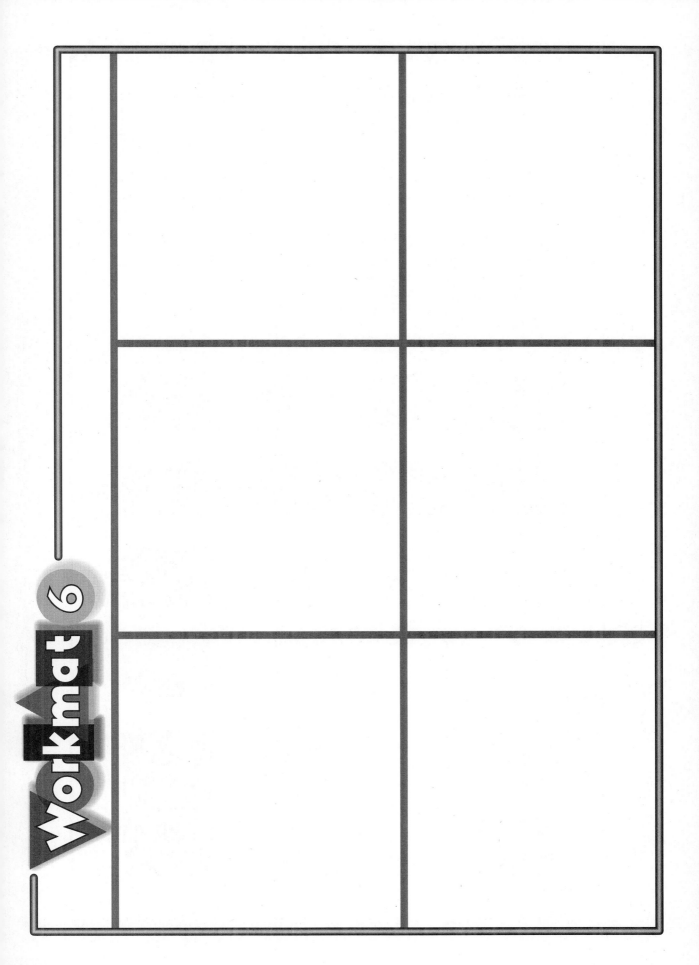

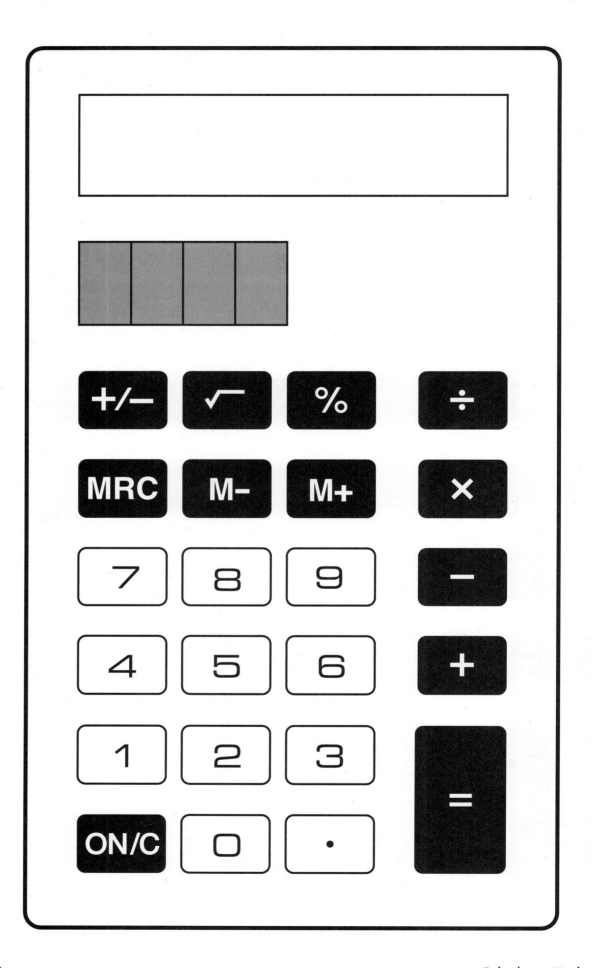

Car Cards

Felt Tree and Apple Pattern

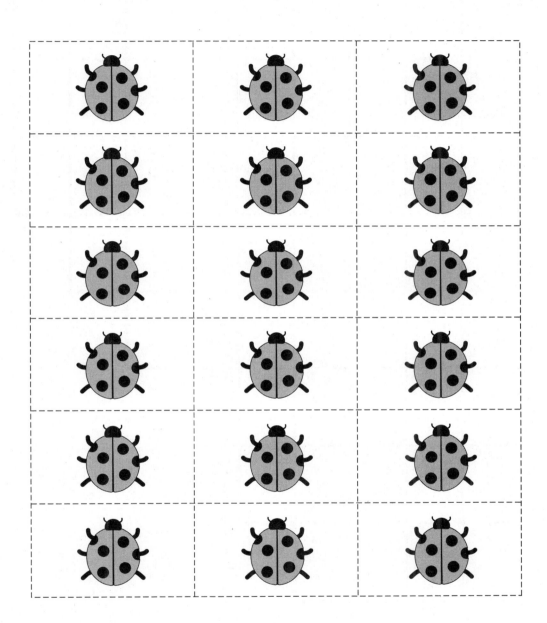

Ladybug Counters

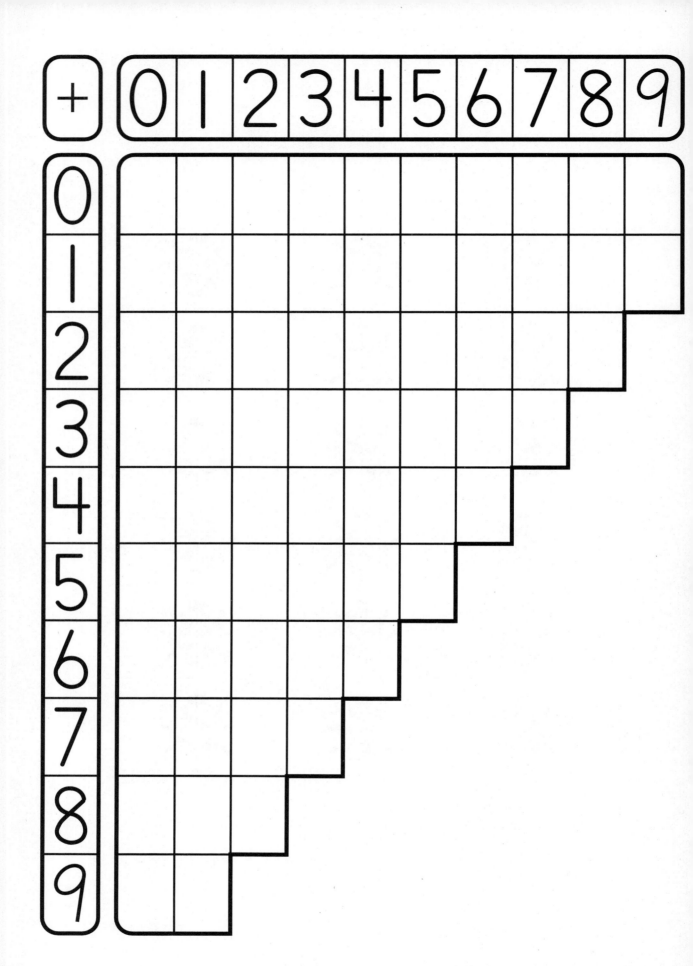

Modified Addition Table

Name _____

Understand

1. Tell the problem in your own words.

_____

_____

_____

2. What do you want to find out?

_____

_____

_____

Plan

3. How will you solve the problem?

_____

_____

_____

Solve

4. Show how you solved the problem.

_____

_____

_____

Look Back

5. How can you check your answer?

_____

_____

_____

VOCABULARY CARDS

Vocabulary Cards are included in this section. A Vocabulary Card is provided for each vocabulary word that is introduced in this grade level.

There are 4 cards on each page. The front of each card contains the word only. The back of the card contains the word and an illustration or sentence that helps children recall the word and its meaning. The chapter in which the word is introduced is noted on the bottom of each card.

Children can keep their cards in a Math Words File—a container such as a file box, shoe box, or zip-top bag. Have them use their Math Words File to review vocabulary terms and to check spelling. Other suggestions for using the Vocabulary Cards appear in the Teacher's Edition on the Follow-Up Strategies and Activities page at the end of each lesson.

plus

in all

equals

sum

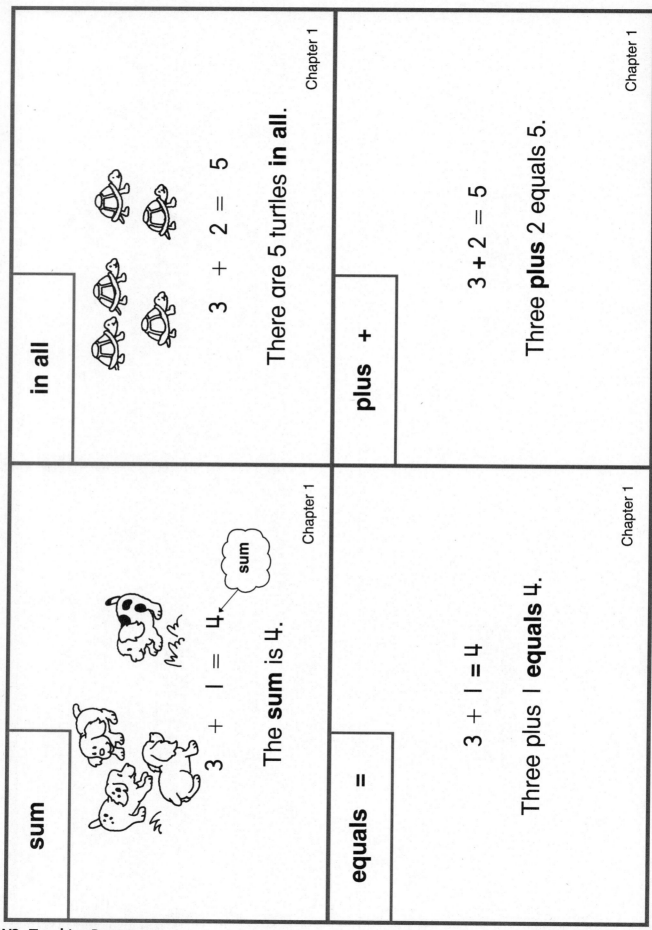

in all

$3 + 2 = 5$

There are 5 turtles **in all**.

plus +

$3 + 2 = 5$

Three **plus** 2 equals 5.

sum

sum

$3 + 1 = 4$

The **sum** is 4.

equals =

$3 + 1 = 4$

Three plus 1 **equals** 4.

addition
sentence

subtraction
sentence

are left

minus

addition sentence

$2 + 2 = 4$

This is an **addition sentence.**

subtraction sentence

$4 - 1 = 3$

This is a **subtraction sentence.**

are left

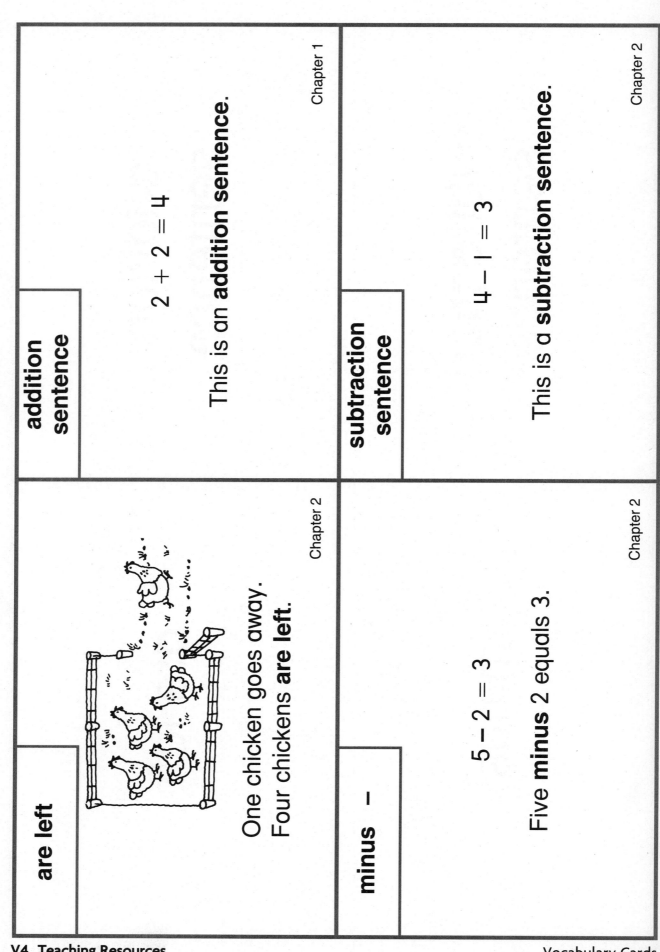

One chicken goes away.
Four chickens **are left.**

minus −

$5 - 2 = 3$

Five **minus** 2 equals 3.

difference

addition combinations

order property

total amount

difference

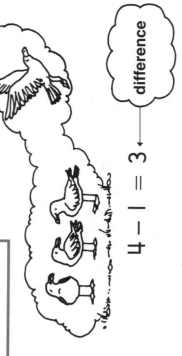

$$4 - 1 = 3$$

difference

The **difference** is 3.

addition combinations

$$5 + 0 \qquad 2 + 3$$
$$4 + 1 \qquad 1 + 4$$
$$3 + 2 \qquad 0 + 5$$

There are 6 **addition combinations** for the number 5.

order property

$$3 + 2 = 5 \quad \text{sum}$$
$$2 + 3 = 5 \quad \text{sum}$$

With the **order property**, you can add numbers in any order and still get the same sum.

total amount

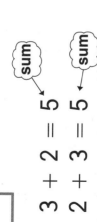

The **total amount** is 2¢.

penny

count on

cent

doubles

penny

1¢ or 1 cent

count on

Count on from 5 to add 5 + 3.

Think 5   Count **6, 7, 8**

5 + 3 = 8

cent

cent sign

1¢

A penny is worth one **cent**.

doubles

4 + 4
5 + 5
6 + 6

When you add the same number twice it's called a **double**.

subtraction combinations

compare

fact family

number line

subtraction combinations

$5 - 5$ $5 - 2$
$5 - 4$ $5 - 1$
$5 - 3$ $5 - 0$

There are *6* **subtraction combinations** for the number 5.

compare

Match the sheep to **compare** the numbers in each row.

fact family

2 + **3** = **5**
3 + 2 = 5
5 − 2 = 3
5 − 3 = 2

A **fact family** uses 3 numbers to make 4 number sentences.

number line

0 1 2 3 4 5 6 7 8 9 10

count back

solid figures

zero

flat face

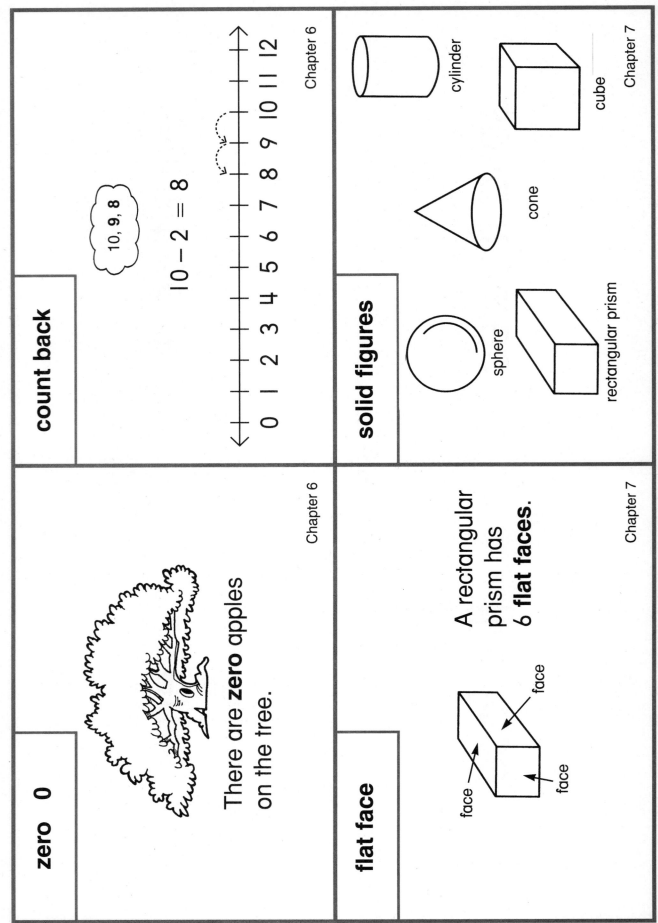

count back

10, **9**, **8**

$10 - 2 = 8$

0 1 2 3 4 5 6 7 8 9 10 11 12

Chapter 6

solid figures

cylinder

cube

cone

sphere

rectangular prism

Chapter 7

zero 0

There are **zero** apples on the tree.

Chapter 6

flat face

A rectangular prism has **6 flat faces.**

face

face

face

Chapter 7

rectangular
prism

cone

sphere

cylinder

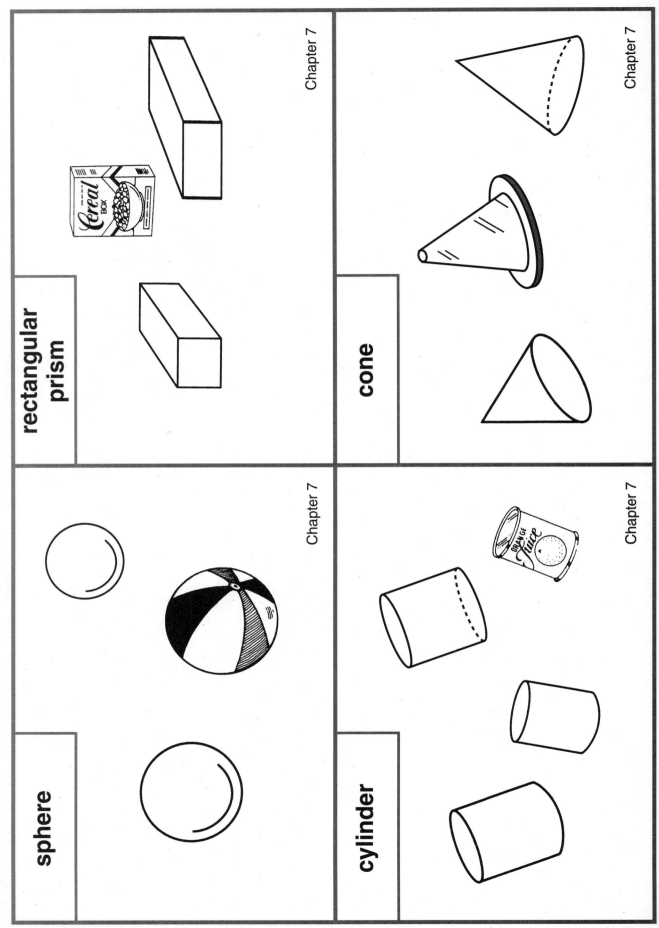

rectangular prism

cone

sphere

cylinder

cube

stack

pyramid

roll

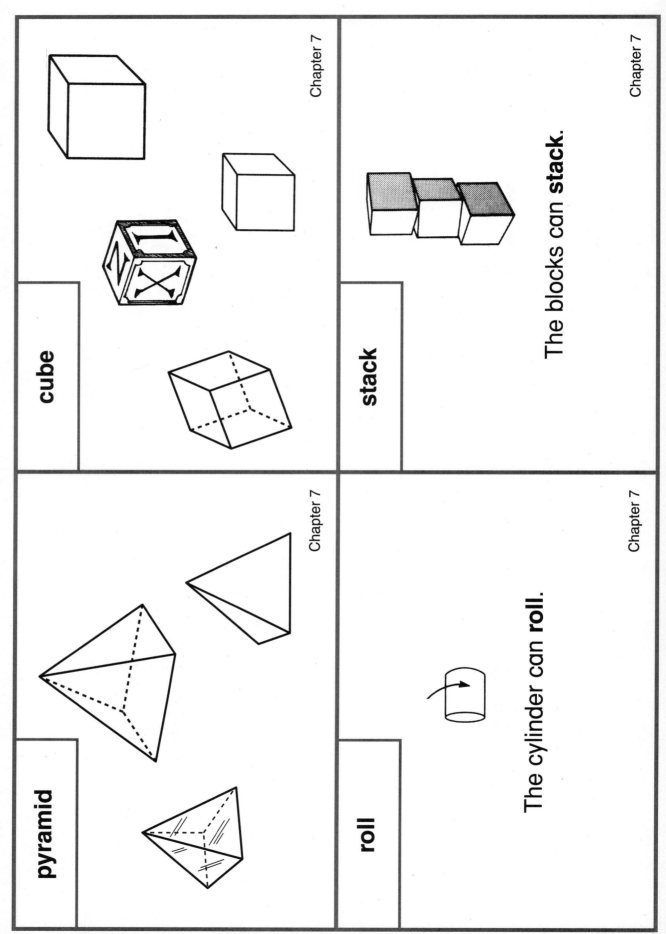

cube

pyramid

stack

The blocks can **stack.**

roll

The cylinder can **roll.**

rectangle

slide

square

circle

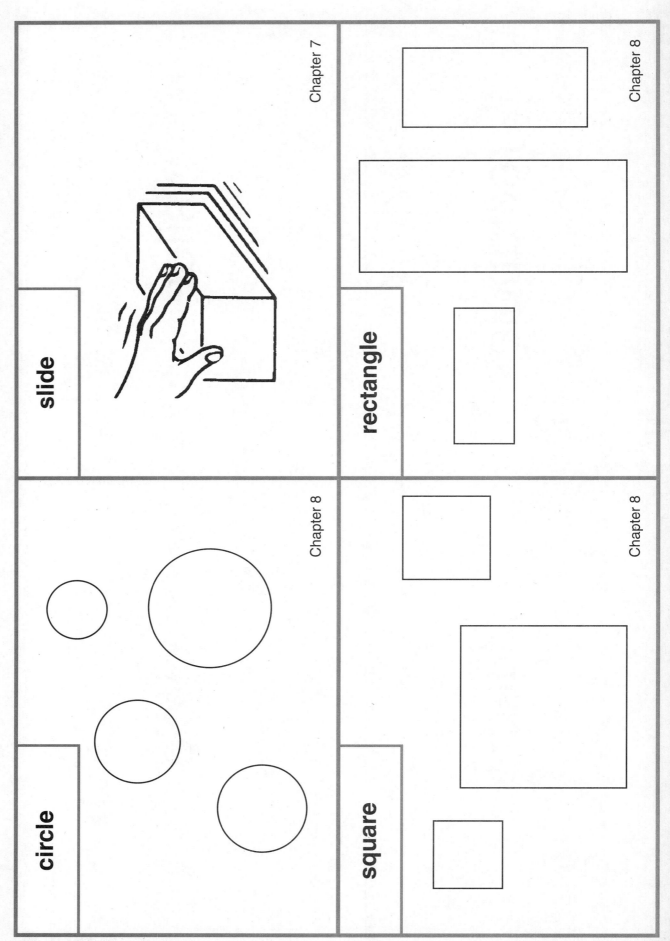

slide

rectangle

circle

square

triangle

corner

side

symmetry

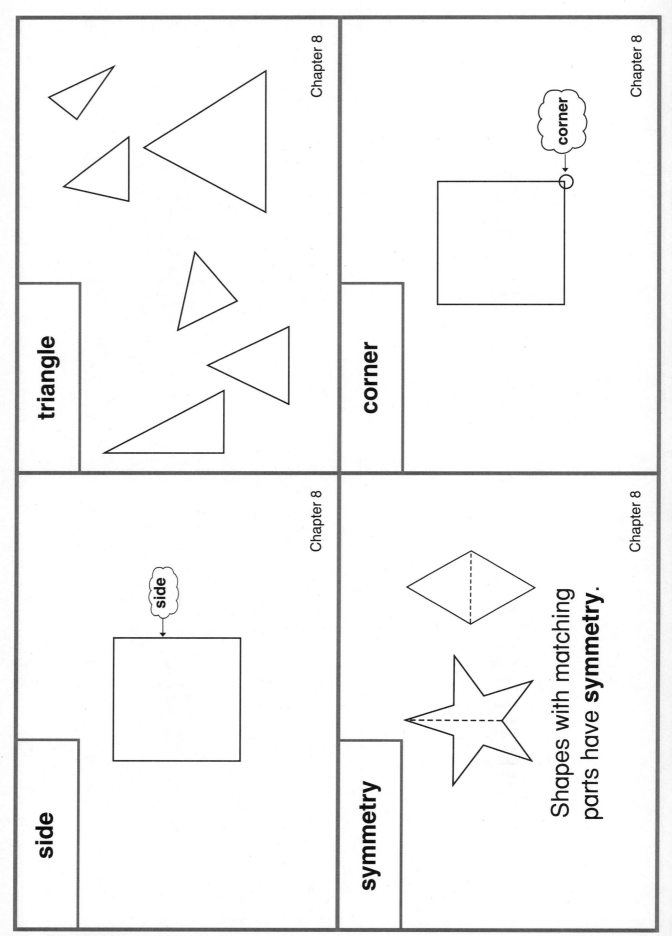

triangle

corner

corner

side

side

symmetry

Shapes with matching parts have **symmetry**.

line of symmetry

closed figures

open figures

inside

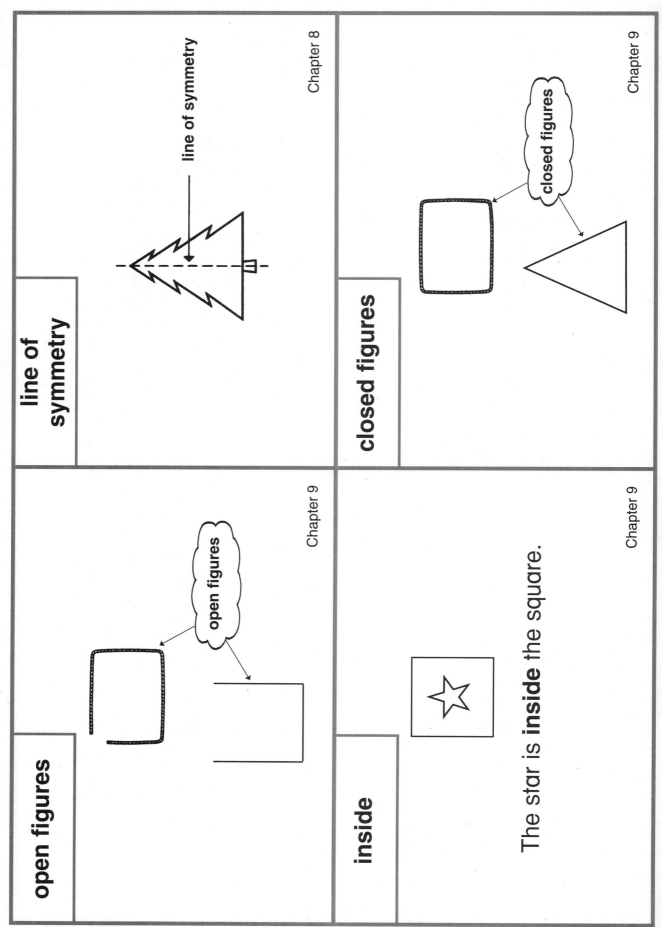

line of
symmetry

line of symmetry

Chapter 8

closed figures

closed figures

Chapter 9

open figures

open figures

Chapter 9

inside

The star is **inside** the square.

Chapter 9

outside

left

on

right

on

The X is **on** the square.

Chapter 9

outside

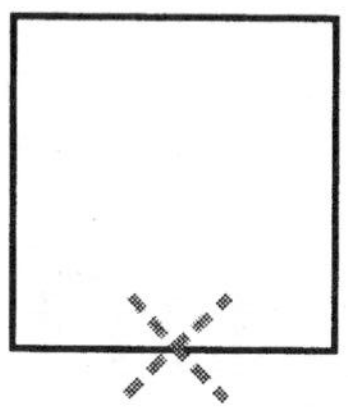

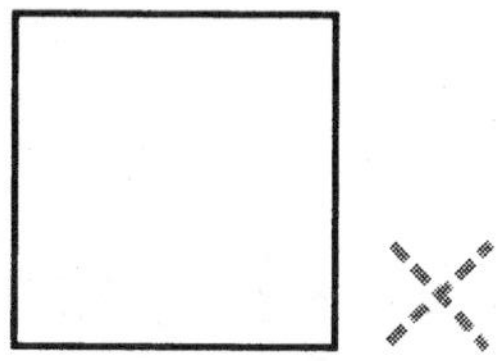

The X is **outside**
of the square.

Chapter 9

right

The flower is to the
right of the path.

Chapter 9

left

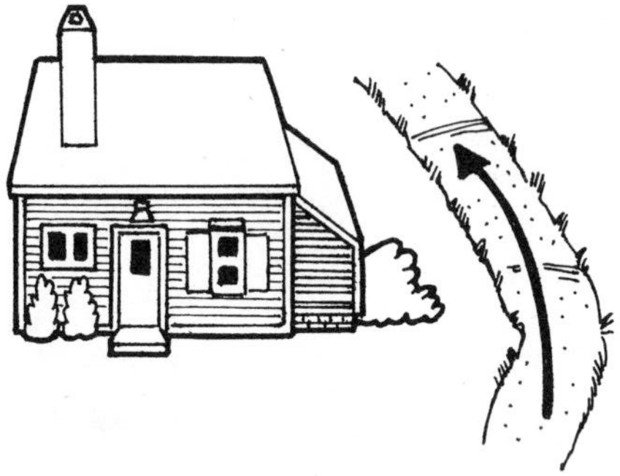

The house is to the
left of the path.

Chapter 9

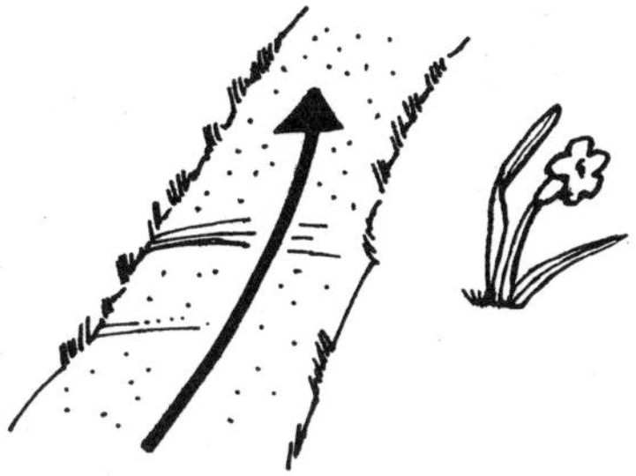

pattern

number
sentence

greater

ten

pattern

△ ○ △ ○ △ ○
△ △ ○ □ ○ ○
△ □ □ ○ □

number sentence

$4 + 2 = 6$
$6 - 3 = 3$

These are **number sentences**.

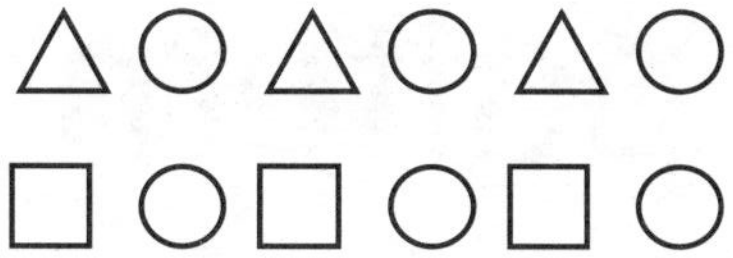

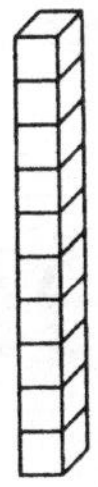

greater

☆ ☆
☆ ☆ ☆
2

☆
☆
5

Five is **greater** than 2.

ten

10

forty

twenty

fifty

thirty

thirty

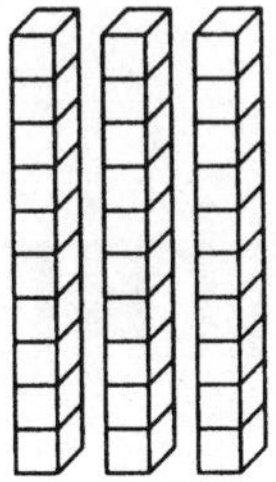

30

Chapter 13

twenty

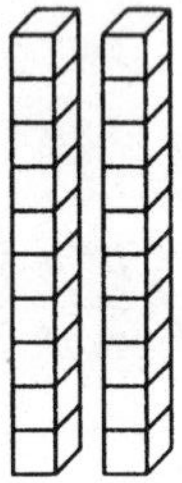

20

Chapter 13

fifty

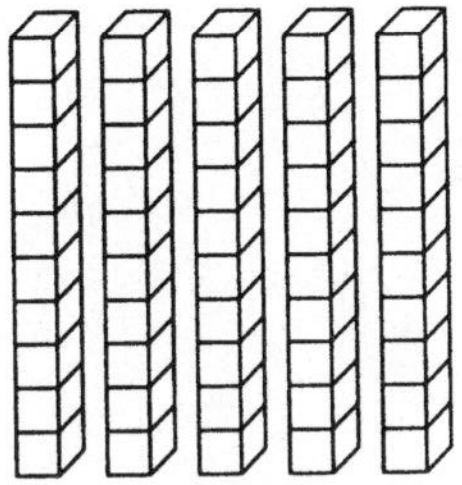

50

Chapter 13

forty

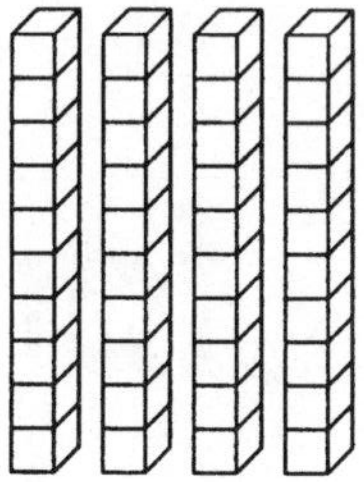

40

Chapter 13

sixty

eighty

seventy

ninety

seventy

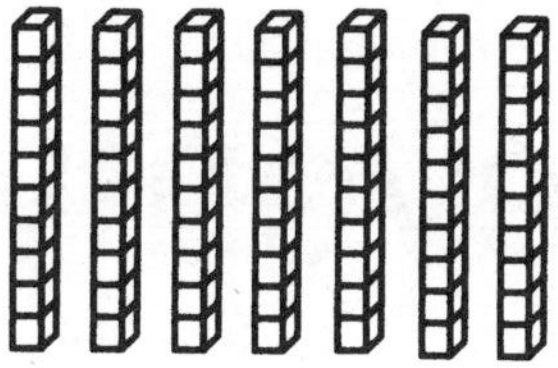

70

Chapter 13

sixty

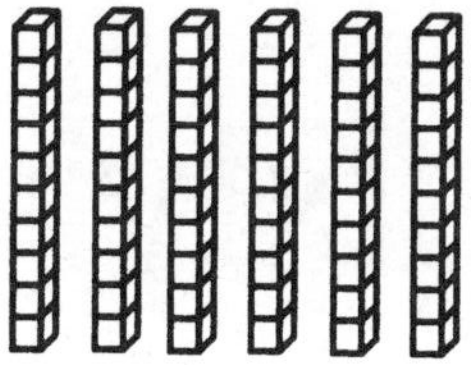

60

Chapter 13

ninety

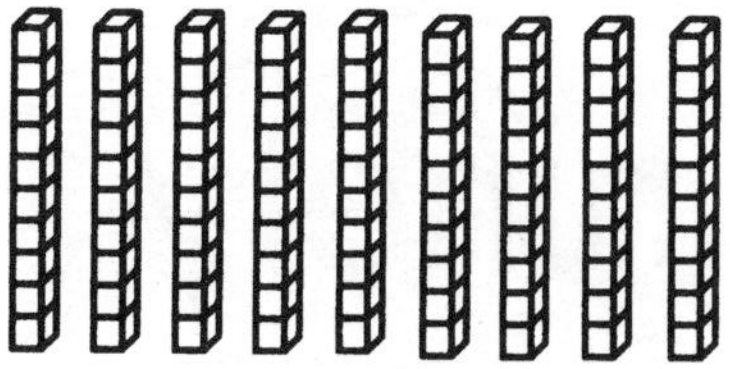

90

Chapter 13

eighty

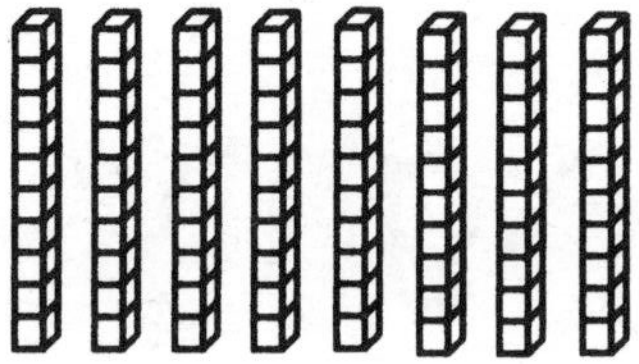

80

Chapter 13

eleven

thirteen

twelve

fourteen

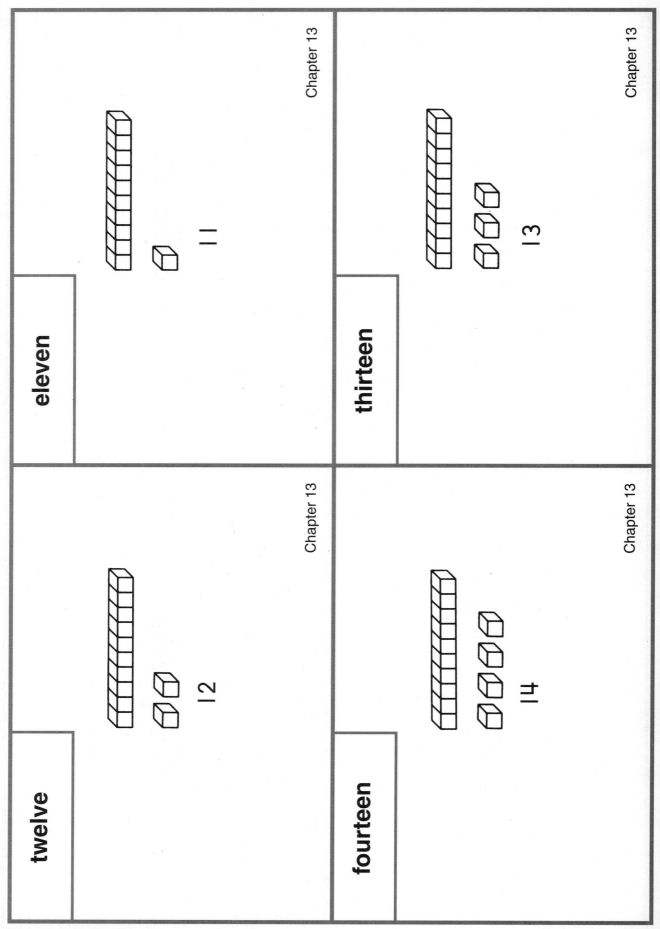

eleven

11

twelve

12

thirteen

13

fourteen

14

seventeen

fifteen

eighteen

sixteen

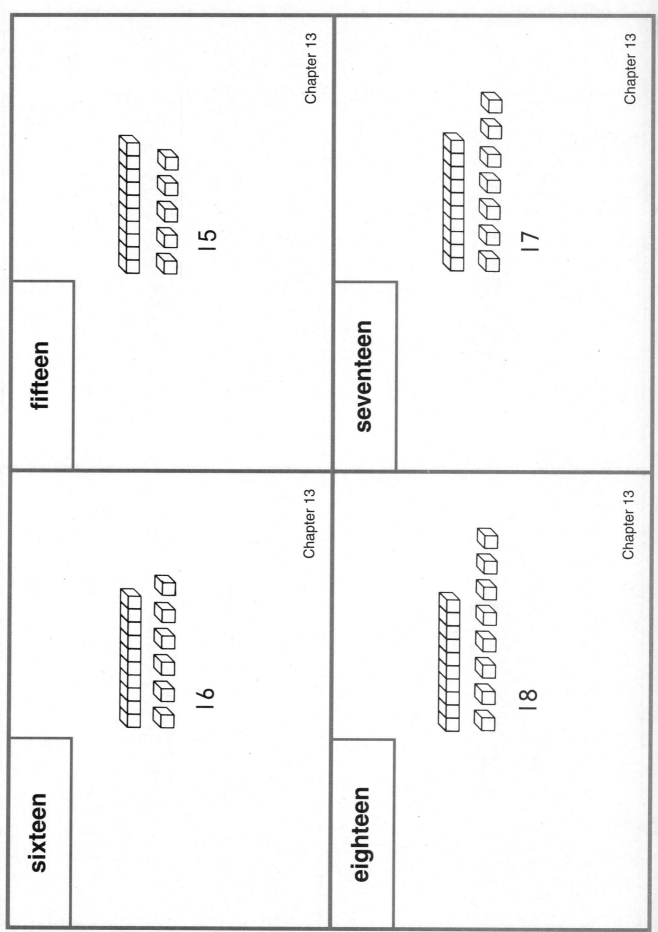

fifteen

15

sixteen

16

seventeen

17

eighteen

18

nineteen

estimate

first
1st

second
2nd

nineteen

19

estimate

Instead of counting the clouds, I **estimate** that there are about 20.

first   1st

The mother is the **first** in line.

second   2nd

The **second** fish is black.

third
3rd

fifth
5th

fourth
4th

sixth
6th

third 3rd

1 2 3 4 5 6 7 8 9 10

The **third** key is black.

fifth 5th

The **fifth** pig is circled.

fourth 4th

The **fourth** bear is circled.

sixth 6th

The **sixth** penny is underlined.

seventh
7th

ninth
9th

eighth
8th

tenth
10th

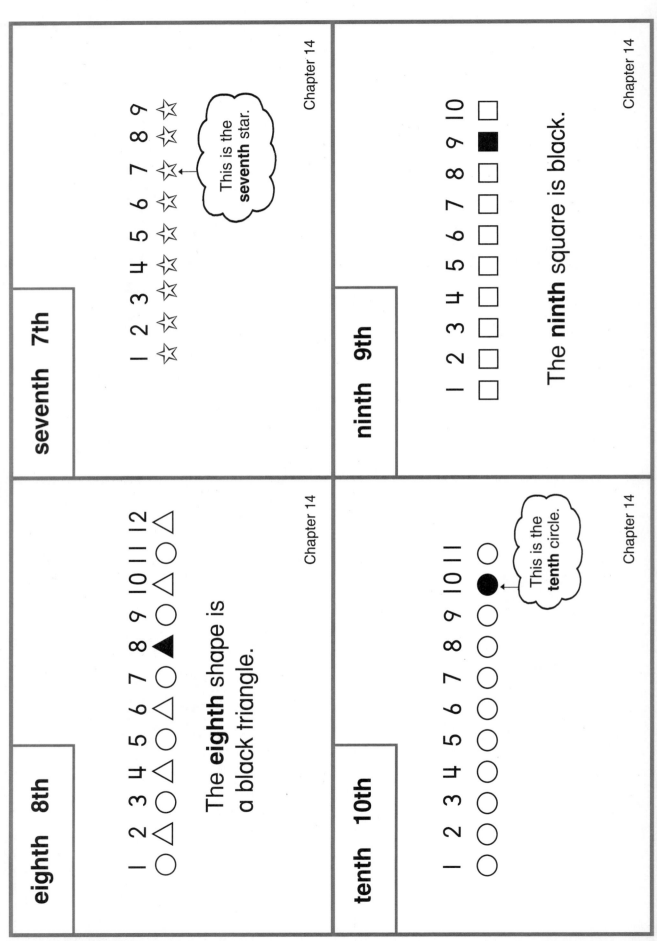

seventh 7th

1 2 3 4 5 6 7 8 9
☆ ☆ ☆ ☆ ☆ ☆ ☆ ☆ ☆

This is the **seventh** star.

ninth 9th

1 2 3 4 5 6 7 8 9 10
□ □ □ □ □ □ □ □ ■ □

The **ninth** square is black.

eighth 8th

1 2 3 4 5 6 7 8 9 10 11 12
○ △ △ ○ △ ○ ○ ▲ ○ ○ △ △

The **eighth** shape is a black triangle.

tenth 10th

1 2 3 4 5 6 7 8 9 10 11
○ ○ ○ ○ ○ ○ ○ ○ ○ ● ○

This is the **tenth** circle.

eleventh
11th

greater than

twelfth
12th

less than

twelfth 12th

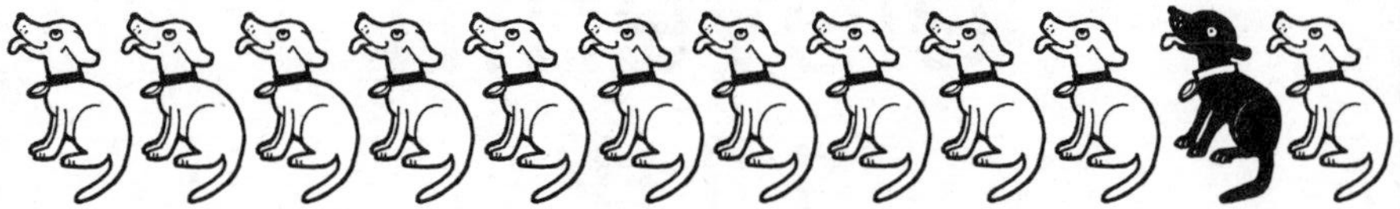

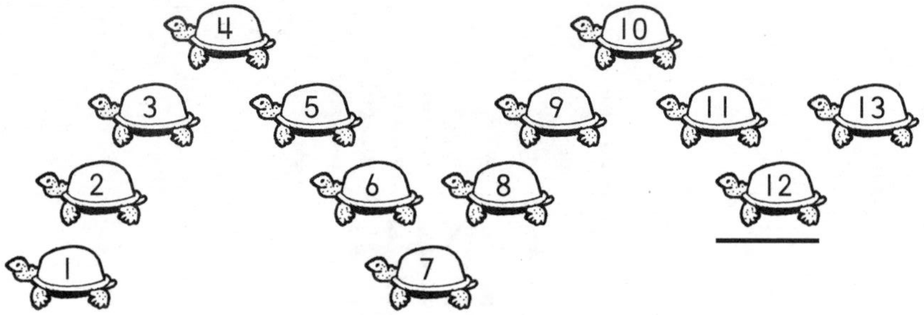

The **twelfth** turtle
is underlined.

Chapter 14

eleventh 11th

The **eleventh** dog is black.

Chapter 14

less than

3 is **less than** 5.

Chapter 14

greater than

5 is **greater than** 3.

Chapter 14

before

between

after

least

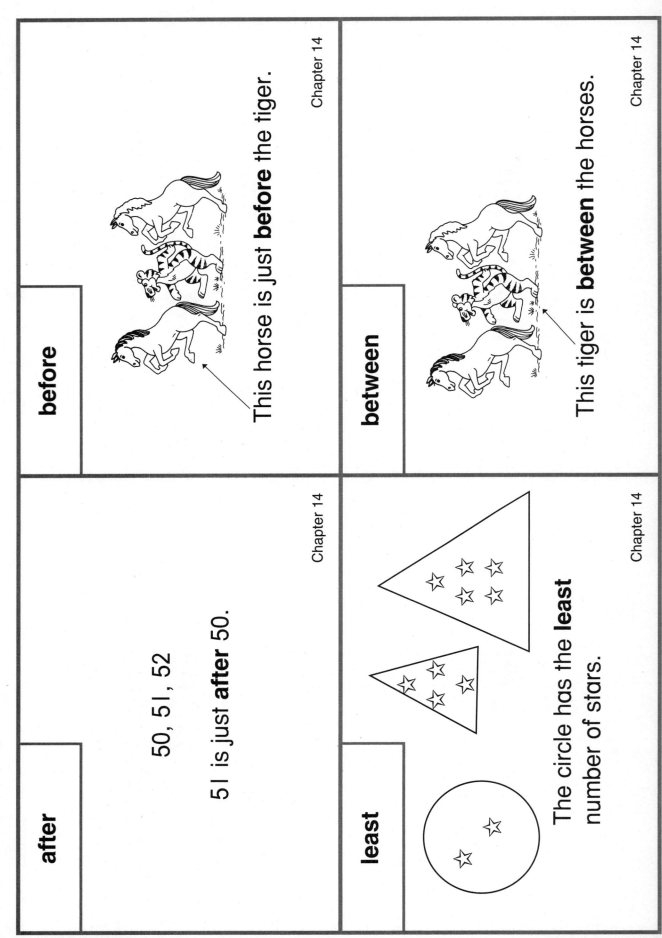

before

This horse is just **before** the tiger.

between

This tiger is **between** the horses.

after

50, 51, 52

51 is just **after** 50.

least

The circle has the **least** number of stars.

greatest

even

odd

nickel

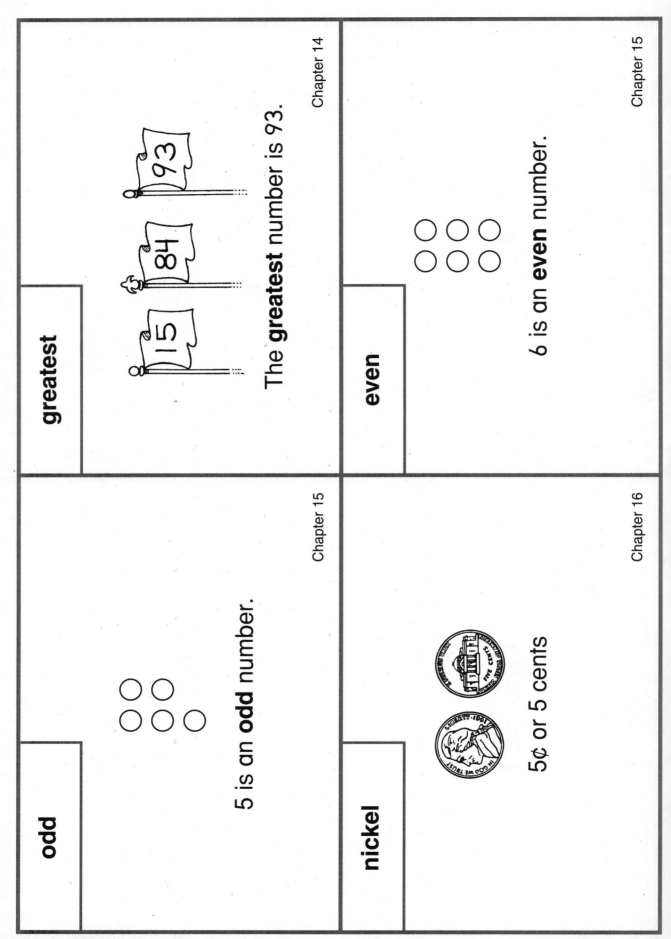

greatest

15 84 93

The **greatest** number is 93.

even

6 is an **even** number.

odd

5 is an **odd** number.

nickel

5¢ or 5 cents

dime

trade

fewest

fewer

dime

10¢ or 10 cents

trade

We can **trade** 10 pennies for a dime.

fewest

Jars Collected

Week 1
Week 2
Week 3

The **fewest** jars were in Week 2.

fewer

There are **fewer** clouds than there are birds.

quarter

February

January

March

quarter

25¢ or 25 cents

February

February is the second month.

January

January is the first month.

March

March is the third month.

June

April

July

May

April

April is the fourth month.

June

June is the sixth month.

May

May is the fifth month.

July

July is the seventh month.

October

August

November

September

August

August is the eighth month.

Chapter 18

September

September is the ninth month.

Chapter 18

October

October is the tenth month.

Chapter 18

November

November is the eleventh month.

Chapter 18

December

days

months

weeks

months

January
March
May
July
September
November

February
April
June
August
October
December

There are twelve **months** in a year.

December

December is the twelfth month.

weeks

There are seven days in a **week**.

days

Sunday
Tuesday
Thursday
Saturday

Monday
Wednesday
Friday

There are seven **days** in a week.

Tuesday

Sunday

Wednesday

Monday

Monday

Monday is the second day of the week.

Sunday

Sunday is the first day of the week.

Wednesday

Wednesday is the fourth day of the week.

Tuesday

Tuesday is the third day of the week.

Saturday

Thursday

date

Friday

Friday

Friday is the sixth day of the week.

Chapter 18

Thursday

Thursday is the fifth day of the week.

Chapter 18

date

Three **dates** have been circled.

Chapter 18

Saturday

Saturday is the seventh day of the week.

Chapter 18

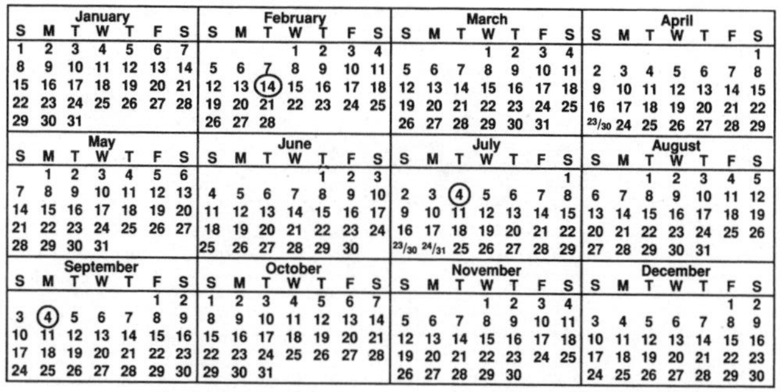

tomorrow

yesterday

morning

today

today

Today is the present day.

Chapter 18

yesterday

Yesterday is the day before today.

Chapter 18

morning

The sun comes up in the **morning**.

Chapter 18

tomorrow

Tomorrow is the day after today.

Chapter 18

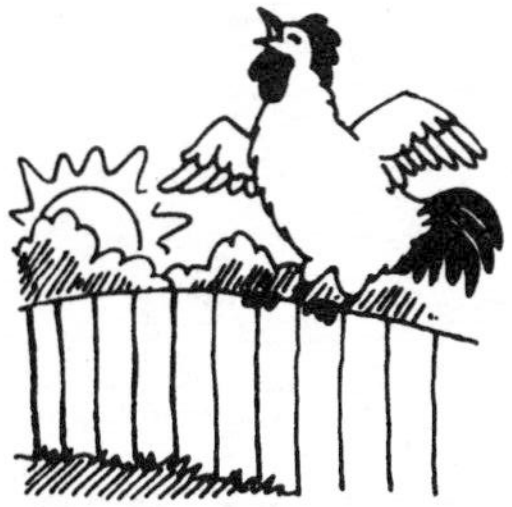

afternoon

longer

evening

shorter

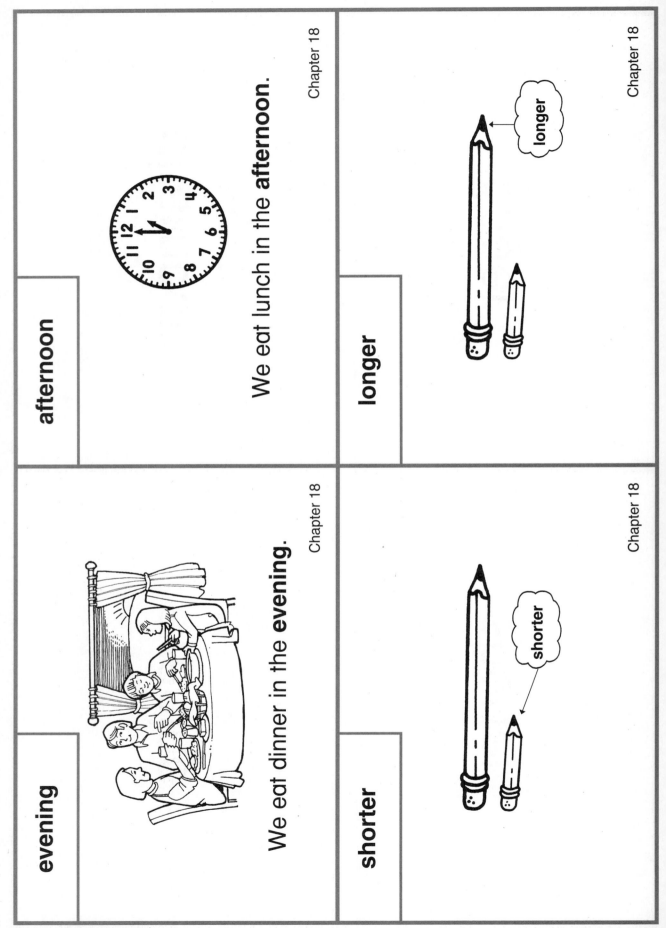

afternoon

We eat lunch in the **afternoon.**

longer

longer

evening

We eat dinner in the **evening.**

shorter

shorter

Vocabulary Cards

minute hand

o'clock

hour hand

hour

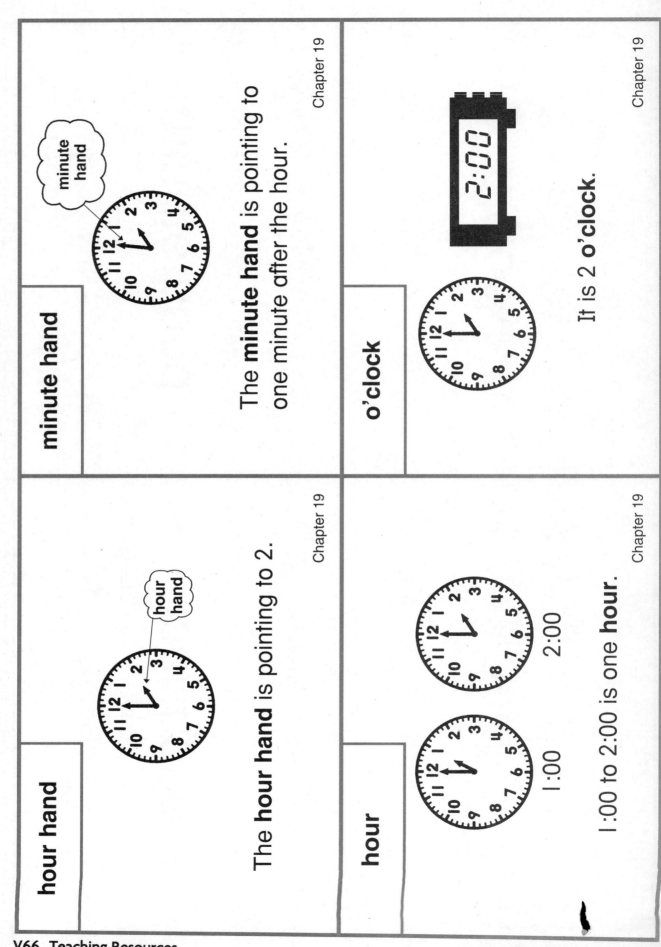

minute hand

minute hand

The **minute hand** is pointing to one minute after the hour.

hour hand

hour hand

The **hour hand** is pointing to 2.

o'clock

2:00

It is 2 o'clock.

hour

1:00   2:00

1:00 to 2:00 is one **hour**.

half hour

thirty minutes after

longest

shortest

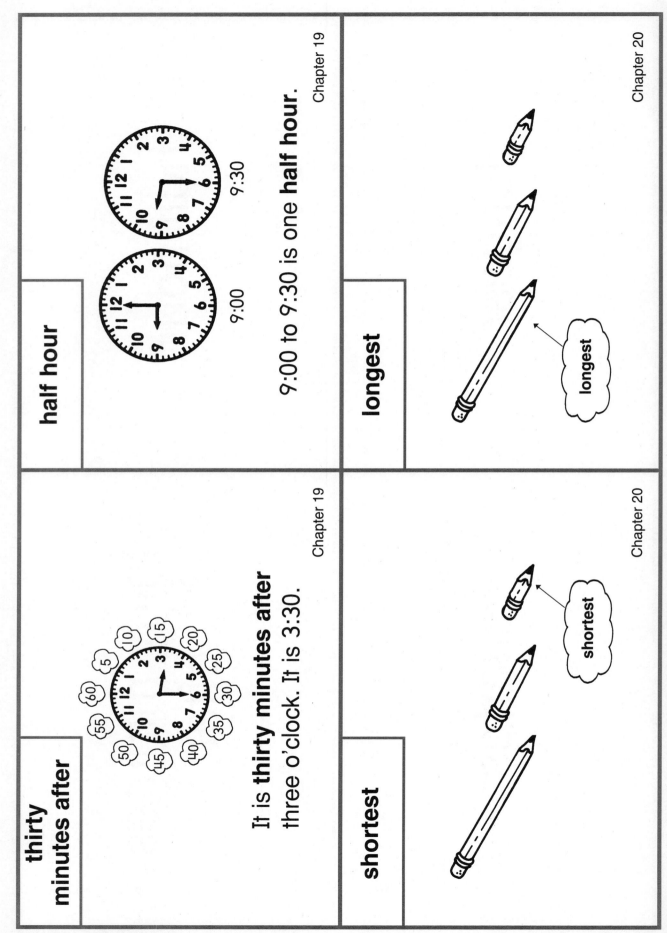

half hour

9:00

9:30

9:00 to 9:30 is one **half hour.**

thirty minutes after

It is **thirty minutes after** three o'clock. It is 3:30.

longest

longest

shortest

shortest

inch

centimeter

heavier

lighter

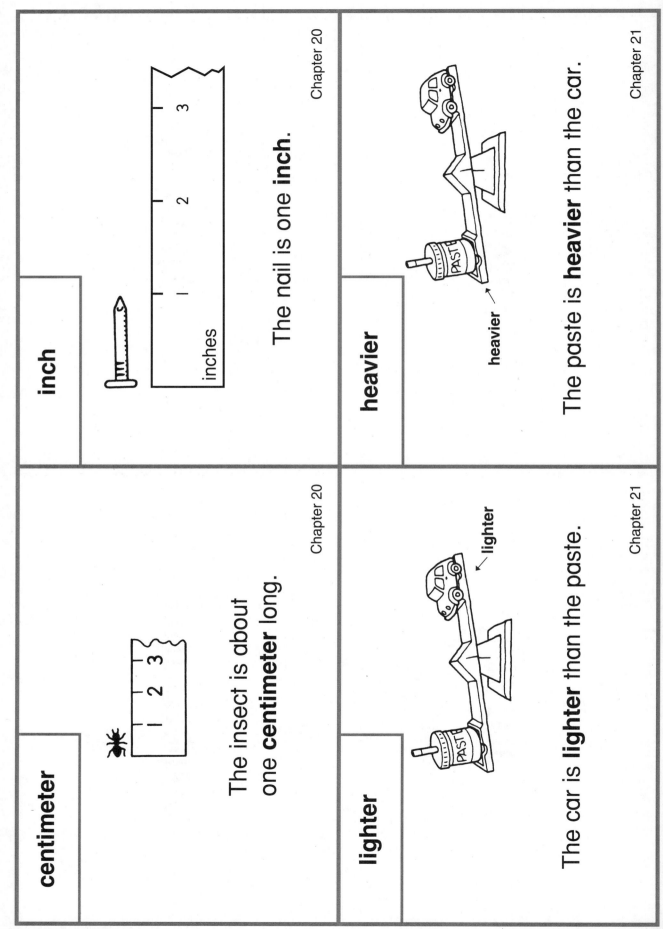

inch

inches

The nail is one **inch**.

centimeter

The insect is about one **centimeter** long.

heavier

heavier

The paste is **heavier** than the car.

lighter

lighter

PASTE

The car is **lighter** than the paste.

Vocabulary Cards

mass

equal parts

measure

halves
one half
$\frac{1}{2}$

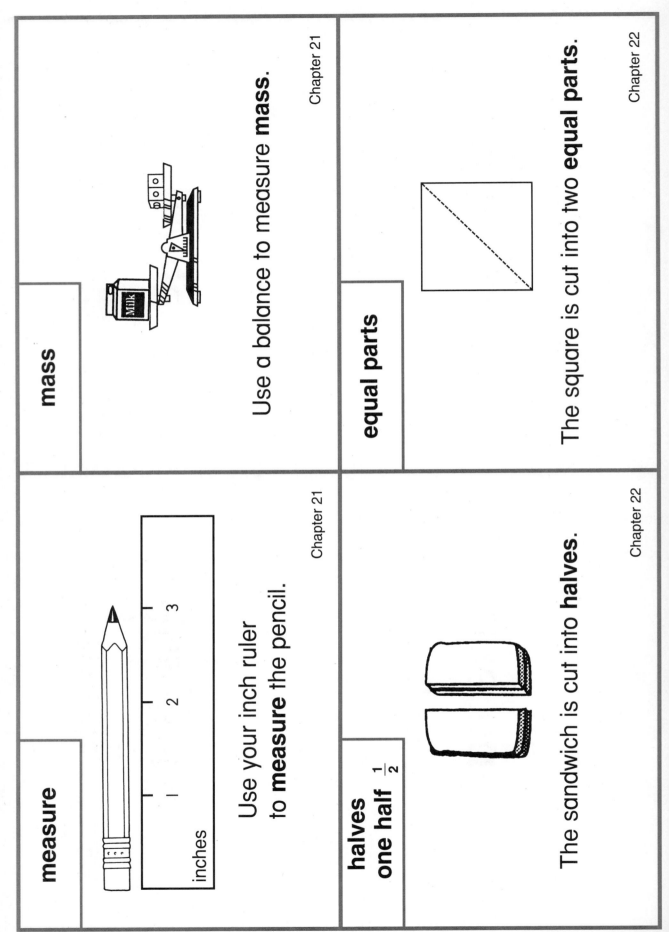

measure

inches

Use your inch ruler
to **measure** the pencil.

mass

Use a balance to measure **mass**.

halves
one half $\frac{1}{2}$

The sandwich is cut into **halves**.

equal parts

The square is cut into two **equal parts**.

fourths
one fourth
$\frac{1}{4}$

sort

thirds
one third
$\frac{1}{3}$

tally marks

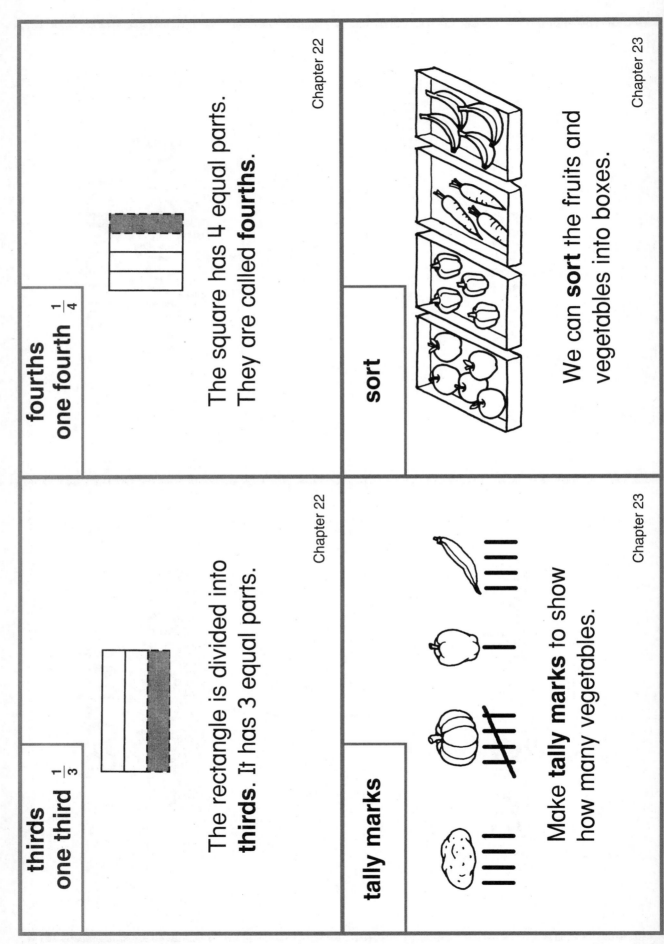

fourths
one fourth $\frac{1}{4}$

The square has 4 equal parts.
They are called **fourths.**

sort

We can **sort** the fruits and
vegetables into boxes.

thirds
one third $\frac{1}{3}$

The rectangle is divided into
thirds. It has 3 equal parts.

tally marks

Make **tally marks** to show
how many vegetables.

certain

most likely

impossible

picture graph

certain

It is **certain** that a circle, a square, or a triangle would be chosen if an object was picked from the box.

most likely

Chapter 24

A circle would **most likely** be chosen from the box.

impossible

It is **impossible** to choose a triangle from the box.

picture graph

Books We Have Read

	0	1	2	3	4	5	6
George							
Mary							
Luis							
Suki							

bar graph

doubles
minus one

doubles
plus one

equal groups

bar graph

Favorite Stories

6						
5						
4						
3						
2						
1						
0						

The **bar graph** shows which story the class liked best.

---

doubles minus one

☆
☆ ☆
☆ ☆
☆ ☆
☆ ☆
☆ ☆

$6 + 6 = 12$
$6 + 5 = 11$

---

doubles plus one

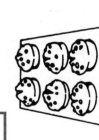

$3 + 3 = 6$
$3 + 4 = 7$

---

equal groups

There are 3 in each group.
They are **equal groups**.